CLEANING SPECIFICATIONS
– their preparation and implementation

By the same author

FUNDAMENTALS OF CARPET MAINTENANCE

AN INTRODUCTION TO CARPET CLEANING

THE IDENTIFICATION OF CARPET FAULTS

DIAGNOSTIC TECHNIQUES FOR THE INVESTIGATION OF CARPET COMPLAINTS

Cleaning Specifications

– their preparation and implementation

Eric M. Brown

Cleaning Research International Ltd.

Copyright © Eric M. Brown 1994

This book is sold subject to the condition that it shall not, by way of trade or otherwise, be lent, re-sold, hired out or otherwise circulated without the publisher's prior consent in any form of binding or cover other than that in which it is published and without a similar condition including this condition being imposed on the subsequent purchaser.

No part of this publication may be reproduced or transmitted in any form or by any means, electronic or mechanical, including photocopying, recording or any information retrieval system without the prior permission in writing from the publisher.

First edition April 1994

ISBN 0 9508446 8 3

Printed and bound in Great Britain by F.M. Repro, 69 Lumb Lane, Liversedge, West Yorkshire, WF15 7NB

Published by Cleaning Research International Ltd., 49 Boroughgate, Otley, West Yorkshire, LS21 1AG, U.K.

PREFACE

For many years *Cleaning Research International* have been involved in the preparation of Cleaning Specifications and their reputation as consultants of international standing is well established. Their specifications are in use at major airports, colleges and universities, television studios, national libraries and museums, atomic power stations and in a multiplicity of offices involved in business and commerce.

This book was begun at the suggestion of one premises manager who wanted a professional specification but recognised that her own building was too small to justify the cost of using independent external consultants. She pointed out that whilst most larger organisations could benefit from *Cleaning Research International*'s input there were many more, smaller ones, who might want to implement the same standards on their own initiative.

Cleaning Specifications – their preparation and implementation, aims to fulfil three objectives: to provide enough detail for the managers of reasonably small buildings to prepare their own specification and to use it in competitive tendering; to illustrate to managers of large institutions the type of specification they ought to implement; and to contribute to the general pool of knowledge and understanding for which the cleaning industry at large is beginning at last, to develop an appetite.

Although the text is presented as the work of one individual, several members of *Cleaning Research International* staff, past and present, deserve much of the credit for the development of procedures described. Of the present staff Jane Devall in particular, supported by Gail Newman, have played an important role in refining and fine tuning the techniques described in this book. Otherwise the basic principles are an amalgam from various sources that have been adapted and extended over the years. The rest, like Topsy, 'just growed'.

E.M. Brown
Leeds
March 1994

AUTHORS NOTE

So that the flow of text is not interrupted by constant interspersion of the many figures used for reference, figures in Chapter 3 *et seq.* are collected together at the end of each chapter. Any stereotyping which appears in the text (contractors as male and operatives as female) is not intended to give offence to either sex but is adopted for simplicity and, in 1994, is a statistically accurate simplification.

The author wishes to thank Motorola European Cellular Subscriber Division. Easter Inch, Bathgate, Scotland for preparation of various floor plans, and permission to use them. The author also wishes to thank Christine Craig for her patience and expertise in the preparation of the proof.

Roger Blackburn, Neptune Stereophonics, Acme Industrial Cleaners Ltd. and Fulcrum Industrial Detergents do not exist. There is no such place as Chorlton-cum-Hardy. Ruritania, Zenda and Rupert of Henzau are products of the imagination of Anthony Hope. Count Dracula is Bram Stoker's.

Contents

1	Introduction	9

Part One

Preparing the Cleaning Specification

2	Case Background	15
3	Preparing the Schedule of Accommodation	22
4	Preparing the Cleaning Schedules – I Core Cleaning Operation	43
5	Preparing the Cleaning Schedules – II Special Operations and Periodics	64
6	Writing the General Notes	80
7	Setting out the Schedule of Conditions	99
8	Defining the Cleaning Terms	124

Part Two

9	**Preparing the Tender Return Forms**	139

Part Three

Estimating the Manpower Requirements

10	Calculating the Manpower Requirements	167
11	Deploying the Workforce	184
12	Costing the Operation	187

1

INTRODUCTION

This book is intended to be a practical guide to all who contract out their cleaning requirements. It also aims to provide technical background and stimulate thought amongst those companies who employ their own cleaning personnel or who offer their services as contractors. Three separate procedures are described:

- The preparation of the cleaning specification

- The drafting of the tender return document

- The estimation of manpower requirements and the evaluation of tenders

A schematic representation of the components of the cleaning specification is shown in Figure 1.1.

PREPARING THE CLEANING SPECIFICATION

A cleaning specification fulfils a number of functions:

1. It forms the basis of a contract between the client and the contractor.

2. It defines the minimum acceptable standards to be attained.

3. It details what cleaning processes need to be carried out and dictates the frequencies with which they should be undertaken.

4. It specifies the areas to be cleaned.

5. It enables an accurate costing of the cleaning operation to be derived.

Additionally an effective specification provides a standard against which the quality of cleaning may be assessed, thereby allowing the implementation of suitable quality control or monitoring procedures.

Figure 1.1: Components of the Cleaning Specification

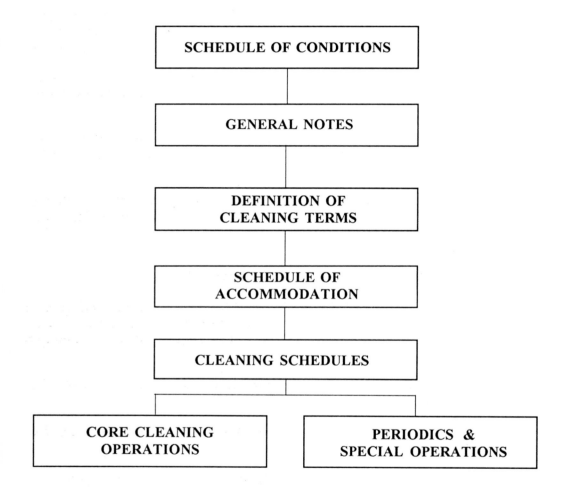

There are five separate components:

- The *Schedule of Conditions* which details the terms under which the contract is to be operated

- *General Notes* which clarify operational aspects of the contract

- The *Definition of Cleaning Terms* providing a glossary of the terms used in the specification and detailing exactly what processes are involved, how they are achieved, and what standards are being sought

- The *Schedule of Accommodation* describing the accommodation to be cleaned, together with floor areas and floor types and classifying this accommodation according to usage

- The *Cleaning Schedules* which specify the actual cleaning processes that are to be undertaken and the frequencies with which they are to be carried out.

When complete, the specification includes each of the above sections although the sequence in which they are drafted does not necessarily follow this same order.

DRAFTING THE TENDER RETURN DOCUMENT

The tender return comprises a series of standard forms to be completed by each competing contractor. This ensures that the information presented can be compared on a like for like basis, thereby simplifying the tender evaluation process. These standard forms:

- Seek to establish the number of hours allocated by the contractor for core cleaning operations

- Seek to establish the number of hours allocated by the contractor for periodic cleaning operations

- Identify the rates of pay for each grade of staff to be employed

- List the equipment and materials to be used

- Specify the deployment of staff

- Detail all costs relevant to fulfilling the requirements of the cleaning specification and therefore indicate the final contract price.

ESTIMATING MANPOWER REQUIREMENTS AND EVALUATING TENDER SUBMISSIONS

Estimating manpower, or *workloading,* is a systematic procedure for the calculation of the time required to clean a building. As such it is a cornerstone in the determination of total costs. In order to undertake the process, the following supporting information or documentation is required:-

- A formal specification detailing tasks to be performed and the frequencies with which they are to be carried out

- A schedule of accommodation listing all of the rooms to be cleaned, the floor areas and the types of floorcovering installed in each

- Information relating to the density of furniture, level of soiling, number of items to be cleaned, levels of occupation and other site-specific details

- Standard times, usually derived from work study data

The evaluation of tenders involves careful comparison of each submission against every other submission and against internally derived estimates of manpower.

For convenience, this book is divided into three separate parts each concerned with one of the three principal procedures outlined above. In order that the student may better understand the stages involved in drafting a comprehensive working specification and ultimately going out to tender, the process is illustrated throughout by means of a case study.

Part One

Preparing the Cleaning Specification

Chapter 2	Case Background
Chapter 3	Preparing the Schedule of Accommodation
Chapter 4	Preparing the Cleaning Schedule – I Core Cleaning Operations
Chapter 5	Preparing the Cleaning Schedule – II Special Operations and Periodics
Chapter 6	Writing the General Notes
Chapter 7	Setting out the Schedule of Conditions
Chapter 8	Defining the Cleaning Terms

2

CASE BACKGROUND

Neptune Stereophonics plc manufacture electronic widgets. They operate from a green field site in Chorlton–cum–Hardy. Their's is a purpose built factory constructed some eight months ago. The building epitomises all the latest thinking in factory design and is thus in keeping with the company's high profile as market leaders in electronic widgets. The 17,000 square metres or so of space is arranged on two floors in such a way that a large proportion of the accommodation is open plan. Roger Blackburn, the Premises Manager can look over the top of the upholstered screen that marks out the territory he calls his office, right across the production area and adjacent warehousing to the tables and chairs of the staff restaurant beyond. A plan of each floor is shown in Figures 3.2 and 3.3.

The building was occupied in phases as staff were transferred progressively from other sites around the country in accordance with a timetable fixed to coincide with the commissioning of different production lines. By the second month 180 personnel had moved in and a cleaning contractor had been appointed to maintain the building to an 'acceptable' standard. At that time, four production lines were operating Monday to Friday on two eight hour shifts, from 0600–1400 and from 1400–2200. Now, six months later, 532 staff are employed; eight production lines are working on three eight hour shifts and weekend working is becoming the norm rather than the exception. Furthermore, despite worldwide recession, the demand for electronic widgets is increasing at such a pace that not only will the factory soon be working at maximum capacity, but another similar facility, almost twice the size of the present one, is to be built next door and will be ready for occupation in fourteen months time.

In order to maintain passable standards of cleanliness, Blackburn by now has started to spend considerable sums of money paying his contractor to supply extra labour at overtime rates. Deteriorating standards in toilets, the restaurant and principal walkways, where the level of cleaning no longer meets the demands of the increased usage, have begun to draw comments from members of staff. Elsewhere certain areas – ventilation grilles, plant rooms, overhead pipeworks, high racking in the 'Goods Outward' warehouse for example – have never been cleaned since the factory was first occupied. To make matters worse VIP visitors from the American parent have taken him to one side to explain that the premises do not reflect the image that the company is trying to present as it opens the doors of its European 'showpiece' to prospective corporate customers.

Roger Blackburn, a Production Engineer by training, fully recognises the importance of specifications. Everything that Neptune Stereophonics manufacture is manufactured to a specification; all of the facilities operate according to a specification. Clearly the cleaning service needs to do so too. The cleaning contractor has offered to prepare one on his behalf and he is sorely tempted to accept. What should he do?

Let us consider some of the problems facing Roger Blackburn and the reasons why they have arisen.

1. The factory is brand new, built only eight months ago. This means there is no real history on which to base a suitable working document. Whatever cleaning needs to be scheduled, the schedule has to be drafted from first principles. (Indeed experience shows that even when a building has been in use for many years, the existing specification is often so poor that the problem still has to be approached from first principles).

2. The extensive open plan layout means that large areas of the factory are visible at all times, under the scrutiny of visitors and staff wherever they may pass by. Thus, inadequate cleaning in just one small section of the building may well downgrade the image of the whole.

3. When the present contractor was first appointed, the total staff numbered approximately 180 and cleaning schedules were fixed accordingly. The agreement with the contractor was that as numbers increased, further labour would be provided on an overtime basis. The consequence of this is that with three times as many staff, 50% more weekday shifts, and increasing occupancy levels at weekends, Blackburn is probably paying more for extra staffing by the cleaning contractor than he is for the agreed scheduled cleaning operation. Clearly a renegotiation of the contract is called for.

4. Some areas have never been cleaned. This is a common error in new buildings with an inadequate cleaning specification. Because the building is new everything starts off in a clean condition. In the case of Neptune Stereophonics for example, the ventilation grilles being brand new can be neglected for several months before a build-up of dust around the vents first becomes noticeable. So too, with the pipework and ducting criss-crossing above the entire production area. Only now is a build-up of dust becoming apparent. Additionally the monoblock paviours around the outside of the building have become contaminated with bird-lime; the carpets are beginning to look grubby; the ovens, sinks, fridges and waste disposal units in the canteen are ready for deep cleaning; and the upholstered screens separating the open plan offices are fingermarked. There are numerous examples of other periodics which have not been specified – and it is far simpler for Blackburn to ask his incumbent contractor to undertake this work, without any real consideration of the cost, than it is to seek separate quotes from a variety of contractors.

Almost a blank cheque for the existing contractor to quote whatever he likes. As a consequence, Blackburn's once 'prudent' budget for cleaning is escalating completely out of control.

No wonder Blackburn is tempted to let his present cleaning contractor prepare a new specification.

Blackburn thought back to the time when he had first appointed Acme Industrial Cleaners Ltd. He was new to the job. Indeed he was new to the company as were most of the other employees on the Chorlton-cum-Hardy site. Furthermore he had never before been responsible for cleaning.

Having no notion how to find a suitable contractor he had called a number of other similar factories that he regularly passed on his way to work and had asked for their recommendations. In this way he had compiled a short-list of four companies, each of whom he invited along to quote.

In every instance he had escorted them around the site and in due course he had received four quotations.

It was a little difficult to decide the relative merits and demerits of each since their views as to how the work should best be carried out differed from contractor to contractor and he felt unable to compare the quotes on a like for like basis. In the end he had settled for Acme who had quoted a fixed price for the core cleaning and had suggested that periodics should be costed as the need arose. He had also arranged a separate window cleaning contract and a third contractor handled sanitary towel disposal and the replenishment of sanitary towel dispensers. This company also supplied the roller towels used in all toilets.

No sooner was the contract set up than an immediate shortfall of effort was identified in connection with the waste management requirements. Each production line generated significant amounts of waste. Components wrapped in polythene, were delivered in cardboard boxes directly to the line. Down the lines, assembly workers expected cleaning staff to remove waste from their work stations. Additionally Neptune Stereophonics, like many other companies were keen to promote themselves as caring for the environment and had begun to implement recycling programmes. From the beginning of last month, cardboard and polythene from production, scrap paper from offices and aluminium cans from the restaurant were being sorted and recycled. As a consequence, Acme now had one full time employee dedicated to the collection, sorting and compacting of waste.

Let us continue to consider Blackburn's problems in detail.

5. At the time he first went out to tender he had no formal specification against which the contractors might quote. Thus the quotes could not be compared on a like for like basis.

   Suppose you went to buy a car and, when asked your requirements by the salesman you answered that you needed a car simply to get from A to B. Suppose he said he had two models to meet your requirement; a Fiat Uno costing £6,000 and a Jaguar XJ6 for £26,000. You would choose the Fiat and save £20,000.

   If however you answered that you wanted a car that would offer spaciousness, comfort, luxury, an automatic gearbox, acceleration and kudos, you would buy the Jaguar.

   Conversely, for fuel economy, ease of parking, cheap maintenance, low price and manoeuvrability in the congested High Street, the Fiat would be your choice.

   In other words as Figure 2.1 illustrates, your selection is determined by your requirements or your *specification*.

   So, using the same analogy suppose you have the choice of two contractors both of whom simply say they will *clean* your premises. One says he will charge you £70,000 per annum. The other will charge £90,000. Again you will save yourself £20,000 by selecting the cheaper of the two.

   But if the levels of cleaning, the frequency of cleaning, the nature of cleaning etc. varies between the two contractors, yet you do not really know what you require, then you cannot make *any* confident choice, (See Figure 2.2).

6. Because of the piecemeal way in which the contract has been put together, routine cleaning, window cleaning and sanitary towel disposal are undertaken by three separate contractors. Furthermore, as we shall see later, the carpet cleaning is also the subject of a fourth contract. There is a strong possibility that because he has four separate contractors on site, Blackburn is paying more than he needs, for to be sure he will be making some contribution to each contractor's overheads. It is quite likely that by placing the contract with only one contractor the sum of all four will be less than he is presently paying. He should at least plan to test this hypothesis.

7. Blackburn's waste management procedures are clearly in need of some rationalisation. We shall see how he handles this problem in due course.

Figure 2.1 Criteria Influencing choice

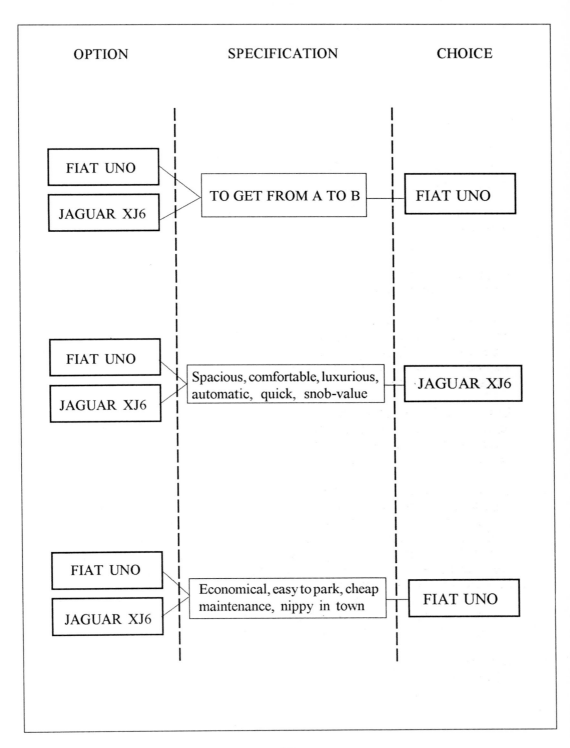

Figure 2.2 Choice without criteria

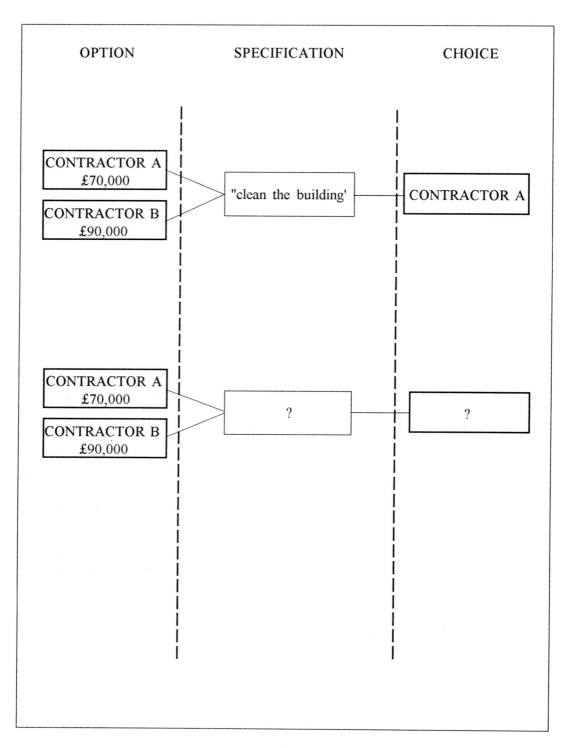

8. Allowing the cleaning contractor to prepare a specification is not a satisfactory solution, for the following reasons:

   i. Some tasks generate a high profit, others generate a low profit.

   There is a clear risk that the contractor will suggest increased frequencies for those tasks which offer the best profit margin. Conversely he has no incentive to suggest those which do not.

   ii. Some tasks require a high capital expenditure. 'Ride-on' scrubber dryers are more expensive than pedestrian operated scrubber dryers for example; Pedestrian scrubber dryers are more expensive than mops, buckets and wet pick-up machines.

   The contractor might thus be tempted to favour labour intensive methods in preference to those which require a high capital outlay at the commencement of the project.

   iii. Allowing the contractor to write the specification allows him to tender the contract to the price he thinks you are willing to pay. It gives you no indication of the price you *ought* to be paying in order to meet the requirements you are seeking.

   iv. The contractor's perception of what levels of cleaning you require may not match your own.

   Thus some areas may be over cleaned whilst others are inadequately cleaned. For example, the contractor may not fully understand the usage of certain rooms; he may be unaware of typical occupancy levels; he may overestimate the importance of some areas of the building whilst underestimating others.

   v. If you plan to go out to competitive tender, allowing the incumbent contractor to prepare the specification gives him an unfair advantage. If you allow each separate contractor submitting a tender to prepare the specification then, as we have already observed, you cannot compare the submissions on a like-for-like basis.

Having carefully considered all of the alternatives Roger Blackburn decided that the time had come to formalise his cleaning contract; to invite competitive tenders for the cleaning of the building according to its present (and forecasted future) requirements; and to prepare the specification himself.

3

PREPARING THE SCHEDULE OF ACCOMMODATION

One of the first tasks when writing a cleaning specification is to carry out a building survey. This is done in order to prepare the **Schedule of Accommodation** and in readiness for drafting the actual *Cleaning Specification*.

The *Schedule of Accommodation* fulfils a number of functions:

- It lists all of the rooms and areas that are to be cleaned
- It classifies the rooms into various categories according to type
- It details the flooring or floorcovering material
- It quantifies the floor area for each room
- It summarises any special requirements pertaining to a particular room or area

An extract from a typical *Schedule of Accommodation* as it applies to Neptune Stereophonics, is shown in Figure 3.1.

In order to complete the *Schedule of Accommodation* it is necessary to have a complete set of floor plans for the entire site. The plans for the ground floor and upper floor of Neptune Stereophonics are shown in Figures 3.2 and 3.3.

Every separate room needs to be designated according to some form of numbering system. It may be that the rooms have been numbered on the doors in such a way that the number indicates both the room and the floor, (as is often the case in hotels where room 211 for example, is room 11 on the second floor). Some other system may be in use. But if none of these apply you will need to number the rooms yourself. Do it systematically on your working copy of the floor plan as you progress around the building conducting your survey.

Next you will need to establish a room classification system. Rooms are classified into categories depending upon their function and the standard of cleaning that is required. For example public or prestige areas such as Reception, conference rooms, the Board room, Directors' offices, will require a different type of service to dining rooms. Emergency exits and staircases may need a lower frequency than main staircases, corridors and lifts.

Decide which categories you will use to classify the different rooms and areas in your own accommodation and prepare a list as an aide memoire. Do not worry if at this stage you are unable to think of every category that you may require. You can add to the list as you complete the building survey.

For convenience assign a two letter code to each category. Some suggestions are shown in Figure 3.4

You are now almost ready to commence your building survey.

However, before you do so there are some other aspects of the specification that might need preliminary consideration before you begin. For example, are there any areas where a daily cleaning frequency is not enough – toilets perhaps, especially in production areas that are working three shifts? Will you need to make special arrangements for deep cleaning the kitchen or the sanitary fittings? Who cleans in food preparation areas – the cleaning contractor or the catering contractor?

With this in mind it is useful, before commencing the building survey, to review briefly a number of questions that may have some bearing on decisions you will make during the survey. Typical questions appear below, at this stage without any explanation. They are considered again, in greater detail in Chapter 6.

1. Assuming that most cleaning will take place out of normal 'office hours' are any 'day cleaning' operations required?

2. Will VDUs, facsimile machines, shredders and other 'hi-tech' equipment be included in the specification?

3. Will telephones be cleaned by the main contractor?

4. Are there special circumstances relating to the cleaning of computer rooms?

5. Are there any rooms with restricted access?

6. Are external areas to be included?

7. Will deep cleaning of sanitary fittings be required?

8. Who will be responsible for deep cleaning of the kitchen?

9. Will carpet cleaning be subject of a different contract?

10. What arrangements will be made for wall cleaning?

11. What arrangements will be made for ceiling cleaning?

12. Are the light diffusers to be cleaned?

13. Is window cleaning to be included or will it be subject of a separate specification? Will the window cleaner be responsible for internal glass partition cleaning?

14. Is there any confidential waste? What are the waste removal requirements? Is waste to be recycled?

15. Should storerooms be cleaned?

Other questions may well apply to specific problems on particular sites and these, together with more general aspects relating to the operation of the contract, are considered in much greater detail in Chapter 6.

Having considered the answers to the above questions fairly superficially at this stage you are now ready to commence the building survey.

In the first room of each category that you visit you will need to make copious notes.

First give the room a number and a title or verbal description – print room, fax room, open plan office, Board room etc. as shown in columns **II** and **III** in Figure 3.1.

Then classify the room according to one of the categories of your room classification system. (Figure 3.4) If you do not have a category to suit the room in which you are standing, create one. Remember your initial list, written before commencing the survey, is only a provisional one. (Column **IV,** Figure 3.1).

Next you will need to make a series of observations about the room each with a specific focus in mind. Make notes on these observations under one of the following headings:

- FLOORS
- WALLS, PARTITIONS, DOORS AND GLASS
- FURNITURE, FIXTURES AND FITTINGS
- REFUSE

FLOORS

Record the nature of the floor or flooring material. It is likely to be one of the following:

Carpet	Vinyl	Linoleum
Marble	Terrazzo	Ceramic tile
Quarry tile	Wood	Concrete
Resin	Safety flooring	Metal (chequer plate)

WALLS, PARTITIONS, DOORS AND GLASS

Are the walls papered or painted? Melamine or fabric? (i.e. can they be wiped down or only vacuumed?). Are there any partitions or screens? If so are they glass or upholstered; or melamine?

Are the doors glass, or wood with glass vision panels?

What is the door 'furniture' (handles, pushplates, kickplates) made from? (Brass, aluminium, chrome, plastic?)

FURNITURE, FIXTURES AND FITTINGS

Any of the following may be included although clearly the list is not exhaustive:

desks, tables, chairs, cupboards, coatracks, filing cabinets, telephones, mirrors, pictures, signage, window sills, ledges, pipes, radiators.

If a particular item is likely to need a particular form of cleaning because of the fabric from which it is made, make a clear note in your record. For example, chairs may be vinyl (which can be scrubbed); upholstered (vacuumed); leather (damp wiped & dried); metal framed (such that the framework needs damp wiping) and so forth.

Make sure you record any special features in the room – a brass rail around the reception desk, a sculpture, a plant feature, a display cabinet, a dado rail, etc.

Some rooms have sinks (a pantry adjacent to a conference room; or a workshop for example) yet may otherwise fall into the category of OFFICES perhaps. Corridors may have a drinking fountain, or an eye wash and emergency shower facility. By the time you have finished your review you will have noted everything that needs to be cleaned.

REFUSE

Note the type of bin. Is it an ashbin, a wastepaper bin, a flip top bin? A Eurocart? Does it have a bin liner?. (If not, should it have?). Note any specialist waste containers including waste for recycling, confidential waste or biohazard waste for example.

Finally* note any aspects of the room that may ultimately appear on your schedule of periodic tasks. Include light fittings, acoustic ceiling tiles, 'hi-tech' equipment, overhead ducts and pipework, etc. Chapter 5 discusses the scheduling of periodic tasks in greater detail.

An extract from Roger Blackburn's working notes is shown in Figure 3.5. Part of his annotated floor plan can be seen in Figure 3.6.

By now the task will appear extremely daunting. It may well have taken fifteen minutes to make these preliminary notes about a single room and you are only at the outset of your building survey. However, let us suppose you began your survey at an entrance to the building (Category EN). The chances are that every other entrance will have similar features. They may not be identical of course but in each subsequent entrance you need now only list exceptions; additional items not appearing on your first entrance inventory. Offices are even simpler. There are fewer items to list (than in a reception for example) and all offices tend to be very similar to one another.

Thus you will spend most of your time in the first room of each category you visit.

The reasons for this will become clear as the cleaning schedules are developed in Chapter 4).

To demonstrate more clearly how this applies in practice, let us look at Figure 3.7 which illustrates Blackburn's working notes for Room 42, the main entrance for employees. We can see that far fewer details appear on this list than there are shown in Figure 3.5. There is a loose laid entrance mat in addition to the coir mat; the floor is vinyl not carpet or granite; the chairs in a small waiting area are upholstered not leather or vinyl; there are some lockers and a time clock. Whatever else may be in that entrance is also to be found in Room 1 (Reception) and therefore does not need to be written down again.

---

* *If you intend to estimate manning levels, this is not the final task Refer to Chapter 10 Calculating the Manpower Requirements.*

Once you have completed your building survey you will have:

- Given every room on the plan a title or description

- Numbered every room in the building according to a logical sequence and marked the number on your floor plan (assuming that the rooms are not already numbered)

- Assigned a category code to each room according to your classification system

- Noted the flooring or floorcovering type

With this information you are able to complete columns **I** to **V** of your *Schedule of Accommodation* as shown in Figure 3.1. You will need to provide an explanation of the abbreviations appearing in column **V**. Figure 3.8 would be suitable.

Additionally you will have:

- Prepared copious notes for subsequent use when drafting the *Cleaning Specification*

Your notes will also allow you to complete column **VII** 'Special Requirements'. Here you may enter a few brief details such as the need to check clean a toilet; the fact that the Pantry (Room 3) is not used every day, and that there is a 'hi-tech' mirror in the Audio Visual Room (Room 7), for which reasons you have decided to classify this room as a Computer Room (CO). The significance of these notes will become clearer later in the text.

One final column remains before the *Schedule of Accommodation* is finalised – Column **VI** which indicates room areas, must be completed. This information is needed for two reasons:

1. It provides the cleaning contractor with data that allows him to prepare an accurate quote.

2. It forms part of the information you will need if you intend to estimate manpower requirements to confirm that the contractor's proposed staffing levels are adequate. (See Chapter 10 – *Calculating the Manpower Requirements*).

If you do not plan to estimate manning levels you may simply require the contractor to make his own measurements of room dimensions which of course he is quite able to do. However if it is your intention to provide a professional specification for the cleaning of your premises it is in your best interests to make sure that your contractor is capable of meeting it and you will need to 'workload' the specification yourself. He cannot meet it if the contract is not adequately staffed.

Room areas are normally determined by measurement and calculation using scale plans although in the absence of scale plans the information can be obtained using a 'Rolatape' type device which is a form of measuring wheel as used by Surveyors. Roger Blackburn has had considerably less difficulty at Neptune Stereophonics since all of the information is available on his computer. Figure 3.9 shows the summary sheet he has prepared for inclusion in his specification to complement the *Schedule of Accommodation*. You will note that he has made an error under the entry for the KITCHEN category where the total dimensions do not equal the sum of the individual dimensions. When processing large amounts of data such as may be necessary for the *Schedule of Accommodation* of a complex site, errors might well arise. For this reason it is customary to disclaim responsibility for the accuracy of the floor areas thereby putting the onus on the contractor to satisfy himself that the detail is correct.

Two clerical tasks remain before this section of the documentation is complete. One is to list all of the rooms according to the classification system you have devised. The second is to colour code plans of the site for easy visual interpretation of the classification. Figure 3.10 illustrates part of the accommodation list prepared by Roger Blackburn for Neptune Stereophonics. Figures 3.11 and 3.12 show his colour coded plans. You will note that several areas are unshaded. These are areas that the contractor is not expected to clean. They include specialist switchgear rooms and some storerooms that do not require cleaning. Two large white areas on the upper floor plan represent areas that do not exist. That is to say, one, on the side of the building, forms part of an atrium; the other is a void to accommodate high racking in the ground floor warehouse. Blackburn has also left the food preparation and dishwash room unshaded because these areas are cleaned by the catering contractor. Nevertheless he has included them on his summary sheet (Figure 3.9) in case he decides to have them cleaned by his cleaning contractor at some future date.

Your *Schedule of Accommodation* is now complete. Figure 3.13 shows how to bring each of the elements on which you have been working, into your final document.

SUMMARY

Before you are able to prepare an accurate cleaning specification you will need to undertake a building survey and prepare a *Schedule of Accommodation*. The *Schedule of Accommodation*:

- Lists all rooms that are to be cleaned

- Classifies them into different categories according to their cleaning needs

- Indicates the nature of the flooring or floorcovering

- Indicates the floor areas of each room

- Highlights any special cleaning requirements

- Provides a set of plans as a pictorial representation of the areas to be cleaned

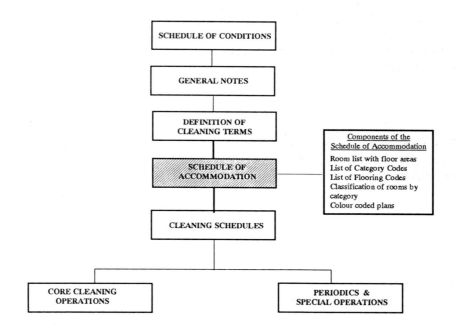

Figure 3.1 Portion of a typical Schedule of Accommodation for Neptune Stereophonics

I	II	III	IV	V	VI	VII
Floor Level	Room No.	Room Title/ Description	Room Classification	Floor Type	Floor Area	Special Require- ments
GD	1	Reception	EN	C/GT/EM	160	
	2	Disabled Toilet	TO	QT	21	Check clean
	3	Pantry	OF	V	9	
	4	Store Room	SR	V	12	
	5	Conference Room	OF	C	124	
	6	Store Room	SR	V	17	
	7	Audio Visual Room	CO	C	23	Hi-tech mirror
	8	Vestibule	CI	V	4	
	9	Storeroom	SR	V	8	
	10	Lift	CI	C	4	
	11	NW Stairwell	CI	V	81	
	12	Corridor	CI	V	45	
	13	Gents Toilet	TO	SF	18	Check clean
	14	Ladies Toilet	TO	SF	16	Check clean
	15	Open Plan Office	OF	C	235	
	16	Corridor	CI	V	45	
	17	Plant Room	PR	V	16	
	18	Gents Toilet	TO	SF	18	Check clean
	19	Ladies Toilet	TO	SF	16	Check clean
	20	PABX Room	PR	V	10	
	21	Seminar Room	OF	C	80	
	22	NE Stairwell	CI	V	74	
	23	Corridor	CI	V	30	
	24	Warehouse	WA	V	910	
	25	Toilet	TO	QT	12	Check clean
	26	Engineering	WO	SF	15	
	27	Corridor	CI	V	30	
	28	SE Stairwell	CI	V	74	
	29	Corridor	CI	V	45	
	30	Main Switchgear	PR	V	80	
	31	Gents Toilet	TO	SF	16	Check clean
	32	Ladies Toilet	TO	SF	18	Check clean

*All areas are in square metres

Figure 3.2 Ground Floor plan for Neptune Stereophonics

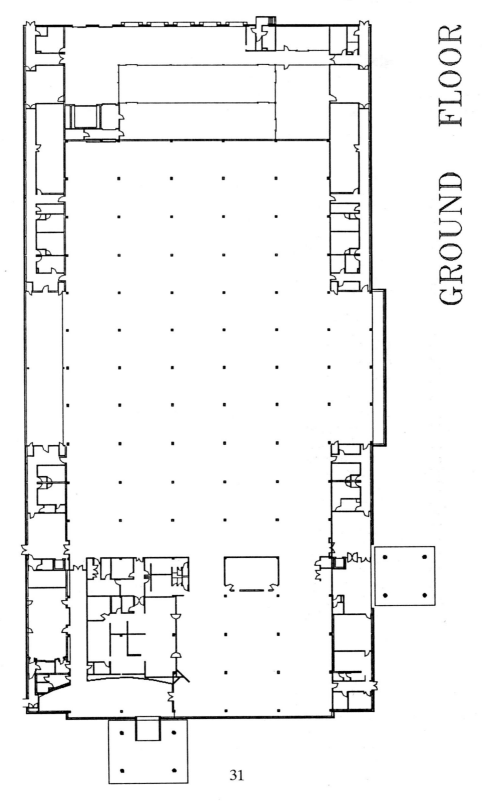

Figure 3.3 First Floor plan for Neptune Stereophonics

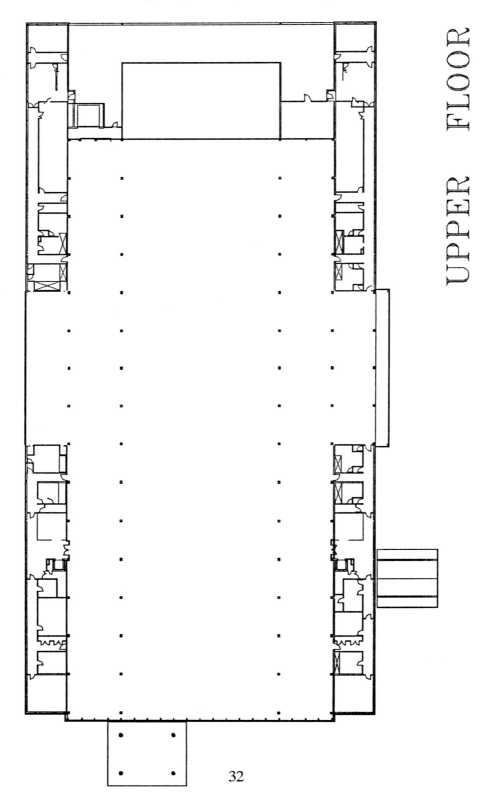

Figure 3.4 Examples of category codes for rooms of different classifications

Classification	Category Code
Entrances	EN
Offices	OF
Circulation	CI
Toilets	TO
Restaurants	RE
Computer Rooms	CO
Production Areas	PA
Warehousing	WA
Staff Rooms	ST
Storerooms	SR
Staircases/Emergency Exits	SC
Medical Rooms	ME
Kitchens, Vending Areas	KI
Plant Rooms	PR
Gymnasia/Sports Rooms	SP
Executive Areas	EX
Laboratories	LA

Figure 3.5 Example of worksheet for Neptune Stereophonics

MAIN RECEPTION: EN   : Room 1

FLOOR: coir mat, carpet, granite

WALLS/PARTITIONS/DRS/GLASS: Painted walls, automatic glass drs, glass screens, wood drs, Al kickplates, push-plates, handles. Vision panels in drs.

FURNITURE/FIXTURES/FITTINGS: Leather chrs, vinyl chrs with Cr frame, glass topped table, fire extng. planter, brass rail round reception desk, phone, sculpture, Co. logo on wall, CCTV, rads, dado rail, pictures, security card reader, display cabinet

REFUSE: Wastebin (Metal) No liner.

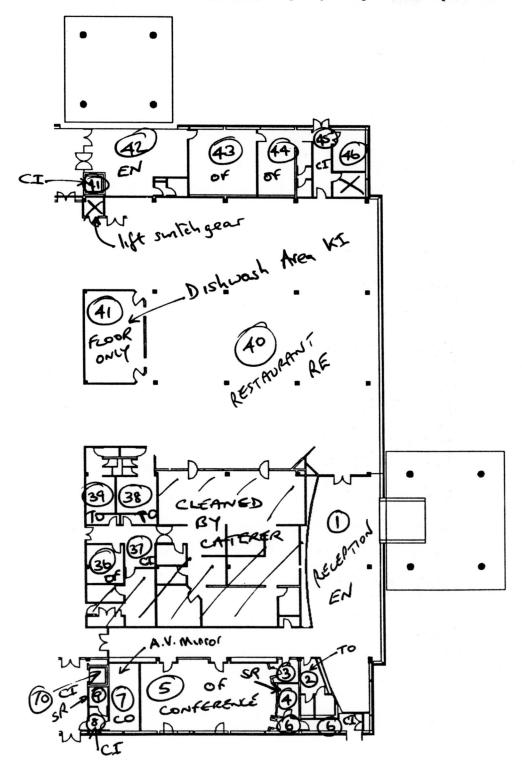

Figure 3.6 Section of annotated ground floor plan for Neptune Stereophonics

Figure 3.7 Worksheet for Room 42 (Employees Entrance) at Neptune Stereophonics

EMPLOYEES ENTRANCE : EN : Room 42

FLOOR: Vinyl tiles, loose laid
ent. mat

WALLS / PARTITIONS / DRS / GLASS :

FURNITURE / FIXTURES / FITTINGS:

uph. chairs
time clock
lockers.

Figure 3.8 Examples of abbreviations for different floor types

FLOOR TYPE	ABBREVIATION
Brick	B
Carpet	C
Ceramic Tile	CT
Concrete	CO
Entrance Matting	EM
Granite Tile	GT
Marble	M
Metal	ME
Quarry Tile	QT
Rubber	R
Safety Floor	SF
Sealed Concrete	SC
Stone	SO
Studded Rubber	SR
Terrazzo	T
Vinyl	V
Wood	W

Figure 3.9 Summary of Areas by category at Neptune Stereophonics

SYMBOL	CATEGORY	CODE	TOTAL (m)	GROUND	UPPER
	OFFICE	OF	3393.71	785.22	2598.47
	CIRCULATION	CI	2934.42	1486.18	1448.27
	ENTRANCE	EN	279.08	279.08	0.00
	TOILETS	TO	389.94	235.62	154.32
	COMPUTER	CO	71.00	21.30	49.70
	STOREROOMS	SR	60.35	26.28	34.07
	RESTAURANT	RE	774.70	774.70	0.00
	KITCHEN	KI	293.07	239.07	0.00
	WAREHOUSE	WA	955.45	955.45	0.00
	PRODUCTION	PA	5935.00	2911.00	3024.00
	WORKSHOPS	WO	15.60	15.60	0.00
	PLANT ROOMS	PR	193.18	151.82	41.36

Figure 3.10 Part of the Summary of Accommodation by category at Neptune Stereophonics

CLASSIFICATION	CATEGORY	ROOM
OFFICE	OF	Offices Atrium Meeting Rooms Conference Room Security Room
CIRCULATION	CI	Corridors Staircases Lifts
ENTRANCES	EN	Reception Employees Entrance
TOILETS	TO	Male & Female Toilets Disabled Toilet Shower Room Changing Rooms
COMPUTER	CO	Computer Room Audio Visual Room

Figure 3.11 Colour Coded Ground Floor Plan for Neptune Stereophonics

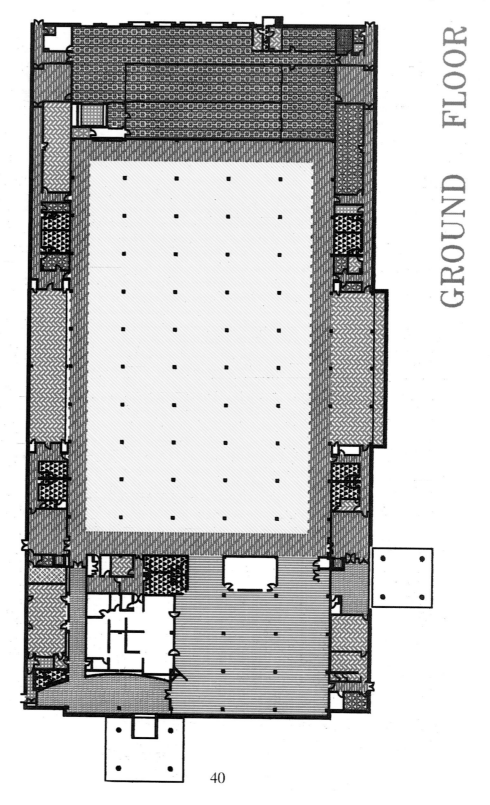

Figure 3.12 Colour Coded First Floor Plan for Neptune Stereophonics

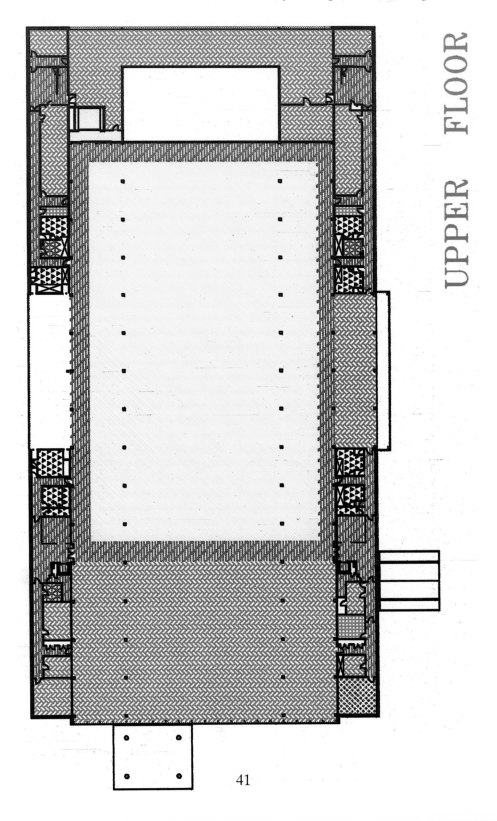

Figure 3.13 Sequence of components of the Schedule of Accommodation

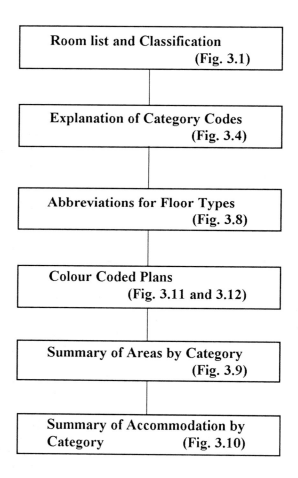

4

PREPARING THE CLEANING SCHEDULES – I CORE CLEANING OPERATIONS

As we have seen in Chapter 3, completion of the building survey allows the *Schedule of Accommodation* to be drafted, rooms to be classified into various categories, and plans to be colour coded. Moreover the comprehensive notes made during the survey are now available for the development of **Cleaning Schedules**. In this chapter we shall be concerned only with the so-called *core cleaning* operations. These are the tasks which are carried out at a frequency of not less than monthly.

A separate schedule will be required for each separate category that has been identified. In the case of Neptune Stereophonics there are eleven different categories, as illustrated by Figure 3.9 (ignoring the food preparation and dishwash areas which have been omitted from the specification since they are to be cleaned by the catering contractor).

A *Cleaning Schedule* for one category of room (OFFICES – OF) at Neptune Stereophonics is shown in Figure 4.1. It can be seen that this schedule provides two fundamental pieces of information:

1. It lists all of the operations that need to be carried out.

2. It specifies the frequencies with which they are to be undertaken.

It may also be seen that, for convenience, these operations are designated under four separate headings: floors; paintwork, walls, partitions and doors; furniture, fixtures and fittings; and refuse. These are the four headings recommended for the collection of data during the building survey as illustrated by Figure 3.5.

Thus, in order to effectively prepare a cleaning schedule for each category of room in your classification, it is necessary to understand the types of operation you may wish to specify, and the frequency with which you may wish to have them carried out.

Definitions of the various cleaning procedures are given in Chapter 8. In this chapter however, we shall consider some of the options that are available and the rationale associated with making each choice. For convenience the options are reviewed under the four headings listed above.

SPECIFYING THE OPERATION

FLOORS

As we have already seen in Chapter 3 a large variety of floor surfaces may be encountered. These include carpet, vinyl, linoleum, marble, terrazzo, ceramic tile, quarry tile, wood, studded rubber, safety floor, concrete, seamless resin, metal.

Furthermore, within each type it is clear that a choice of maintenance techniques is available. Some of these techniques form part of the core cleaning schedule, whilst others are carried out only periodically. Figure 4.2 summarises the more important core cleaning operations that may be specified. (On page 70 of Chapter 5 typical periodic procedures are reviewed).

Selecting the correct procedure presumes a certain level of understanding of basic cleaning techniques. However without this knowledge it is still possible to proceed provided that you are willing to spend some time verifying that what you plan to do is indeed appropriate. There are a number of ways in which you might do this.

For example, you can ask your present cleaning contractor what techniques are currently being used. You can ask the manufacturers of the flooring materials or floorcoverings used on your premises to see what they recommend. You can ask your janitorial supplier for suggestions. You can read around the subject in a suitable textbook, trade magazine, or product technical information sheet. To make the task a little easier, the following guidance notes may help.

Carpet Cleaning Techniques

Remove Stains and Spillages: A procedure which requires no particular explanation except that it should be borne in mind that it is a discretionary instruction to the cleaner. That is to say if there are no stains or spillages to be removed then the operation will not be carried out. Implicit in the instruction is that the cleaner will actually check to see if indeed there are any stains or spillages to be removed.

Vacuum: Vacuum cleaning is the principal core cleaning operation concerned with carpets. The task may be further subdivided into more specific techniques with differing levels of performance. Typical subdivisions are:

Spot-vacuum: Primarily a litter pick-up operation usually involving a canister cleaner with a wand attachment.

Vacuum traffic lanes: Only those areas which are frequently walked upon are vacuumed. These include doorways, aisles between desks in open plan areas, the centre of corridors, areas around vending machines, photocopiers, time clocks etc.

Vacuum edges and corners: Performed to remove any build up of dust in places not easily accessible to an upright machine or the flat head of a canister machine. A special nozzle may be attached to carry out this operation. Alternatively, and more commonly, the head of the canister machine may be removed and the aluminium tube that forms the wand may be used as a nozzle instead.

Fully vacuum: All areas of the carpet are vacuum cleaned.

Smooth Floor Cleaning Techniques

Vacuum cleaning may be carried out on smooth floors in certain instances. For example it is sometimes used in hospital wards to remove gross amounts of fluff before mopping or spray cleaning. However, the most common techniques are:

Dust control mop: Using flat mops, impregnated flat mops or scissor mops, dust control mopping is necessary to remove fluff and loose debris before damp mopping, wet mopping or spray cleaning is carried out.

Damp mop: A variety of damp mopping techniques exist and normally involve wetting the mop and then squeezing it as dry as possible before use. Squeezing may involve an elaborate wringer system, a mechanical lever system, or simple twisting of the mop head in a perforated funnel. In some situations the procedure *spot mop* may be specified. This technique is a derivative of damp mopping but is concerned only with the removal of spots stains and random marks.

Spray clean: This involves spraying a fine dispersion of water detergent or buffable detergent onto the floor and then passing over it with a rotating brush or pad. The procedure is intended to cut through soils that may perhaps be too ingrained to be removed by simple mopping.

Spray burnish: Similar to spray cleaning, spray burnishing is carried out on floors to which a buffable or semi-buffable polish has been applied. The process removes superficial soil together with a layer of polish in which soil has become ingrained. A dispersion of buffable detergent or polish sprayed onto the floor during the operation replaces the polish 'cut' away in the cleaning process. Spray burnishing normally leaves a higher gloss than spray cleaning.

Buff: Buffing is carried out to remove scuffs and to generate a shine on a polished floor. Ideally such machines are fitted with a vacuum and skirt to minimise the creation of airborne dust. No water or detergent is involved, and the machine generally operates at fairly low rotational speeds.

Ultra speed burnish: Ultra speed burnishing is performed using a machine which rotates typically at 2000 rpm. Like buffing, no water or detergent is involved. In theory the high levels of friction developed between the pad and the floor causes the polish to change thixotropically and acquire a shine even greater than that which can be created by buffing.

Neutral scrub: As the name implies scrubbing is primarily used where the floor is particularly dirty and more mechanical action than can be provided by mopping or spray cleaning is required. It is achieved using a rotary machine fitted with a reservoir to hold the cleaning solution which is then fed directly through the brush (or floor pad) onto the floor. Scrubbing is usually followed by a *wet pick-up* operation to remove excess water from the floor. Combined scrubber-dryer machines are available to increase the productivity of this otherwise time consuming two stage process. The preface *neutral* refers to the pH of the cleaning agent and ensures that acid or alkaline chemicals which might otherwise damage certain floors, are not used.

In the illustration given in Figure 4.1 you will see that two types of floor have been scheduled for cleaning in the *OFFICE* category. At Neptune Stereophonics, some offices have vinyl floors; others are carpeted. This is not uncommon in many premises. In factories for example it is usual for offices associated with production and which are situated next to the production area, to have vinyl flooring, whilst offices for management, sales, marketing and other support functions will be carpeted. But even in a city centre tower block, offices may be found with vinyl floors in a building which otherwise has large areas of carpet. 'Security' offices, Despatcher's rooms, fax rooms and mail rooms for example may demand all the cleaning operations that a normal office requires (and therefore do not justify a separate category in your classification system), yet may have a different floorcovering to most of the other offices in the building. It is far simpler to subdivide the *FLOORS* section of the *Cleaning Schedule* as has been done in Figure 4.1 than to prepare a new category that differs from the *OFFICE* category only in the way that the floors are maintained.

In some areas you may find that there is a mix of floorcovering types within that one area. Take a look at Roger Blackburn's notes for the Reception at Neptune Stereophonics for example. (Figure 3.5) It may be seen that there is an entrance mat, carpet and granite tiles.

Thus, based upon this single room sample the *Cleaning Schedule* for the category *ENTRANCES* at Neptune Stereophonics will subdivide the floors into:-

    1.1   Carpet
    1.2   Granite
    1.3   Entrance Mat

In fact, because Room 42, which is the employees entrance on Blackburn's annotated floor plan (Figure 3.6) is also classified in the same category as Room 1, then it too will have the same cleaning schedule. However, unlike Reception, the employees entrance has a vinyl floor and so the FLOORS component of the *Cleaning Schedule* for ENTRANCES will actually have four subdivisions as shown in Figure 4.3.

PAINTWORK, WALLS, PARTITIONS AND DOORS

Many operations concerning paintwork, walls and partitions are carried out as part of the *Periodic Schedule*. Nevertheless, there are some which are core cleaning tasks. Figure 4.4 shows a selection for a typical office location. (Note that these are more comprehensive than those specified for Neptune Stereophonics in Figure 4.1).

Considerably less understanding of basic cleaning techniques is required in order to specify these operations and those concerned with furniture, fixtures and fittings than in the case of floor cleaning. Procedures you may wish to specify include the following:

Spot wipe/spot clean: Random marks such as finger marks, grease marks and splashes are removed by this operation. Just as *remove stains and spillages* is a discretionary instruction for carpets this too is discretionary for paintwork, walls, partitions and doors. If there are no marks the procedure is not carried out.

The distinction between wiping and cleaning is only subtle and the two terms are almost interchangeable. However whilst it may be understood that marks on doors or walls can either be *spot wiped* or *spot cleaned*, clearly the instruction *spot wipe upholstered screens* would not be valid since wiping is not a term consistent with the removal of marks from fabric. Thus *spot wipe* tends to be used as an instruction when merely passing a damp cloth (moistened with detergent) over the mark will suffice, whilst *spot clean* suggests that something more is involved. By the same logic glass tends to be spot cleaned rather than spot wiped. Simple wiping is unlikely to remove greasy finger prints from a glass door for example, – greater effort is required.

If the entire surface is to be cleaned rather than just random marks that may be present you would specify *damp wipe* or *clean*, omitting the word *spot* from the instruction.

Remember too that wiping procedures normally involve use of a detergent. The cloth may either be dipped into a solution of the detergent and wrung out, or it may be moistened by means of a spray bottle containing a detergent solution.

Damp dust: Whereas wiping is carried out to remove marks that may be a little stubborn, dusting is intended to remove only superficial airborne dust. You should note however, that dry dusting (q.v.) should only be carried out in special circumstances (when the surface may be damaged by water such as may be the case with antique furniture for example). Dusting using a dry yellow cloth, as is typical in a domestic situation, is not acceptable. Dry dusting merely disturbs the dust on the surface so that it becomes airborne again only to settle once more on the surface some minutes, or even hours later. Dusting should be carried out using a cloth moistened with neutral detergent solution from a spray bottle, or using an impregnated cloth or dust control device. Never permit a cleaning operative to use feather dusters.

FURNITURE, FIXTURES AND FITTINGS

Figure 4.5 shows a fairly comprehensive schedule for cleaning furniture, fixtures and fittings in office accommodation. Clearly the operations which are specified – *dust, damp wipe, clean*, etc. are the same as those for paintwork, walls, partitions and glass.

Do not be concerned that you may miss some item that needs cleaning. A 'catch-all' statement in the *General Notes* section of the specification (see Chapter 6, page 84) will ensure that nothing is omitted.

Some special operations such as *sanitise telephones fortnightly,* or *anti-static wipe VDU screens monthly,* may also be included in the specification for furniture, fixtures and fittings although in the case of Neptune Stereophonics a separate schedule for VDUs, facsimile machines, shredders and photocopiers, was prepared. This is shown in detail in Figure 5.2.

REFUSE

As companies become increasingly keen to demonstrate that they have an environmentally 'correct' attitude to waste disposal, the scope of operations under this heading is tending to become more diverse. At its simplest, a schedule such as is shown in Figure 4.6 may be all that is required. In more complex situations it may be advisable to prepare a separate *Waste Management Schedule*. This will especially be the case in manufacturing environments where, in order to comply with the Environmental Protection Act 1990 and Duty of Care Regulations under which wastes must be documented from 'cradle to grave' some organisations are delegating greater involvement to their cleaning Contractor.

One category – TOILETS, requires an additional subdivision on the *Cleaning Schedule*. An illustration is shown in Figure 4.7. It may be seen however that separate elements of the toilet cleaning operation need not be itemised. For example the direction *clean WC* presumes that the standard procedure for cleaning a WC will be carried out in its entirety. This involves flushing the cistern and forcing the fresh water back beyond the S bend with a toilet brush; spraying internal surfaces with cleaning agent which is allowed to dwell; wiping or washing both sides of the seat, the cistern, all pipework and handles and the outside of the pedestal; rinsing and wiping dry as necessary; finally cleaning the inside of the bowl and flushing one more time.

Clean showers, which is normally included in the TOILET classification, presumes that the curtain or screen will be wiped in addition to the standard procedure of cleaning the tray. TOILET operations also include replenishment of supplies (which will have been itemised in the *General Notes* section of the specification document).

SELECTING THE FREQUENCY

The cost of meeting the specification will be highly dependent upon the frequencies with which each task is carried out. The two examples which follow show Roger Blackburn's calculations for cleaning OFFICE areas at Neptune Stereophonics.

Roger Blackburn's total office accommodation occupied 3393.71 square metres. Of this all offices on the ground floor (785.22 square metres) were carpeted whilst all those on the first floor were vinyl (2598.47 square metres).

His preliminary thoughts for scheduling the daily duties in the carpeted offices included:

- *spot vacuum floor*
- *dust tops of desks, tables and cupboards*
- *empty wastebins and replace liner (if required)*

Amongst his weekly tasks he was considering:

- *full vacuum floor, clean edges and corners*
- *spot wipe paintwork*
- *dust framework of furniture, skirtings, pictures, window sills*
- *damp wipe wastebins*

However, not satisfied that these frequencies would give him the standards he sought, he considered a second option. Maybe, he should increase vacuum cleaning to become a daily full vacuum operation instead, leaving only edges and corners to be cleaned weekly. Maybe the furniture should also be damp dusted daily rather than weekly.

His alternative schedule was amended as follows:–

- *full vacuum floor*
- *dust framework of furniture, skirtings, pictures, window sills*
- *empty wastebins and replace liner (if required)*

all on a daily basis, and weekly:

- *vacuum edges and corners*
- *spot wipe paintwork, walls*
- *damp wipe wastebins.*

What are the cost implications of Roger Blackburn's two options?

In order to answer this question it is necessary to know typical times used to calculate the labour requirement to undertake each of these tasks. The values Blackburn used are shown in Figure 4.8.

Blackburn's offices are scheduled to be cleaned on a five days per week basis. Thus to find the total time *per annum* to clean 100 square metes of carpeted offices he must multiply his daily times by 260 (which is the number of days service per annum for a five day per week cleaning operation); and he must multiply his weekly times by 52. The cost of cleaning 100 square metres is then simply a multiple of the hours worked and the hourly rate of pay. Blackburn has in fact 785.22 square metres of carpeted office.

His calculations are shown in Figures 4.9 and 4.10 from which we can see that Option 2 is more expensive than Option 1 by 820.72R/449.15R i.e. by a factor of 1.82 or 82%, simply as a result of increasing frequencies of vacuuming and dusting.

The scheduling of vinyl floor cleaning operations may be even more critical. Suppose for example that Blackburn had considered the following alternatives for cleaning the vinyl floor in the offices:

Option one:
- dust control mop   daily
- damp mop   weekly
- spray clean   monthly

Option two:
- dust control mop   daily
- damp mop   daily
- spray clean   weekly

If the standard times for dust control mopping, damp mopping and spray cleaning 100 square metres of vinyl floor are 0.09, 0.23 and 0.35 hours respectively, option one would cost:

$$£[(0.09 \times 260) + (0.23 \times 52) + (0.35 \times 12)] \times 25.9847^1 \times R$$

for Blackburn's 2598.47 square metres of vinyl floored office at an hourly rate of £R i.e. £1027.95R, whilst option two would cost:

$$£[(0.09 \times 260) + (0.23 \times 260) + (0.35 \times 52)] \times 25.9847 \times R$$

which is £2634.84R or more than twice as much (2.56 times)

Try to determine what frequencies and techniques are currently being used and decide whether they are yielding an acceptable standard. If they are, ask yourself whether you can actually cut back a little. If you can, then certain frequencies you specify in your schedules should be lower than present practice. If on the other hand, you feel there is a need to increase frequencies in order to achieve better results, calculate the cost of doing so in the manner demonstrated above.

Your scheduling of core cleaning operations is thus, almost complete. Before you begin to develop your periodic requirements however, there is one final aspect to consider – the *Day Cleaning Schedule*.

SPECIFYING DAY CLEANING OPERATIONS

Whether or not you require a day cleaning service will depend upon the time of day that the core cleaning operations are carried out and upon the number of people using the building. At Neptune Stereophonics, Roger Blackburn has not yet decided exactly when he requires his cleaners to be on site although he feels sure it will be outside normal office hours. (The pros and cons of selecting different options are discussed in detail in Chapter 6). If this is likely to be the case then for sure some cleaners will be needed for the hours of 0900 to 1700 or even during the main shift periods of 0600–2200. However with something like 400 people in the building over these periods, the toilets will need attention, the staff dining room may require cleaning, and spillages and other emergencies may occur.

A typical *Day Cleaning Schedule* is likely to specify the number of staff you require to be on site and the hours during which the cover is to be provided. It will also indicate the precise duties you expect to be carried out.

An example is shown in Figure 4.11. As you will see in the illustration, these staff are described as janitors rather than cleaners and at Neptune Stereophonics there is also a 'handyman' function. In some cases it may be desirable to specify that one of the day cleaning staff should be male and the other female so that mens' and womens' toilets can be serviced by a cleaner of the appropriate gender. This is not essential of course and in any case, your cleaning contractor is prevented by law from practising sex discrimination when hiring staff. Irrespective of this restriction the 'handyman' function is best filled by a male since a degree of lifting – perhaps in the way of furniture removal, is often involved.

Aspects of the day cleaning service (such as 'check clean toilets') may be highlighted in column **VII** of the *Schedule of Accommodation* as illustrated by Figure 3.1

Think carefully through all of your requirements for a day cleaning service. Once complete you have finished drafting your schedule for core cleaning operations.

SUMMARY

The scheduling of core cleaning operations involves establishing what cleaning procedures are to be adopted and what frequencies are required. Additionally it involves the specification of a day cleaning service.

A separate schedule is drafted for each category of room and procedures for dealing with floors; paintwork, walls, partitions and doors; furniture, fixtures and fittings; and refuse, are itemised.

Some understanding of basic cleaning techniques is required and the cost implications of setting frequencies should be taken into account.

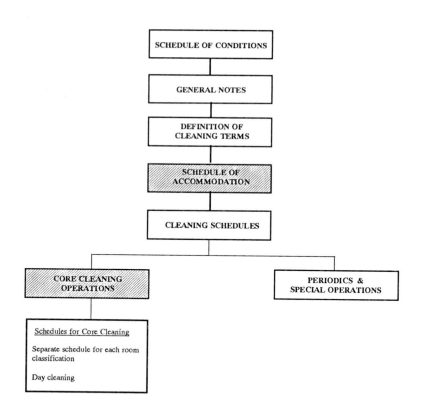

Figure 4.1 Typical Cleaning Schedule for OFFICE areas at Neptune Stereophonics

CLEANING SCHEDULE – OFFICES   CATEGORY CODE OF

Cleaning Requirements	Frequency
1.0 FLOORS	
1.1 Carpet	
remove stains and spillages	daily
spot vacuum	daily
full vacuum	weekly
1.2 Vinyl	
dust control mop	daily
damp mop	daily
spray clean	weekly
2.0 PAINTWORK, WALLS, PARTITIONS AND DOORS	
spot wipe doors	weekly
clean and polish dry glass in doors	weekly
spot wipe paintwork, partitions, pillars	weekly
damp dust doors including frame and furniture	monthly
3.0 FURNITURE, FIXTURES, FITTINGS	
damp dust tops of all furniture	daily
damp wipe telephones	weekly
damp dust framework of furniture, fixtures and fittings, polish dry any metalwork	weekly
damp dust ledges, shelves, skirting boards, window sills, pipes, radiators, wall fixtures and fittings below 2m	weekly
spot clean and remove debris from upholstered chairs	weekly
vacuum upholstered chairs	monthly
4.0 REFUSE	
empty wastebins and dispose of rubbish, replace liner if required	daily
damp wipe internal and external surfaces of wastebins	monthly

Figure 4.2 Typical core cleaning operations for different flooring/floorcovering types

Flooring/Floorcovering	Operation
Carpet	remove stains and spillages spot vacuum vacuum traffic lanes vacuum edges and corners full vacuum
Smooths (vinyl, linoleum, marble, terrazzo etc.)	dust control mop damp mop spray clean burnish ultra speed burnish spray neutral scrub buff

Figure 4.3 Floor cleaning Schedule for ENTRANCE areas at Neptune Stereophonics

CLEANING SCHEDULE – ENTRANCES   **CATEGORY CODE EN**

Cleaning Requirements	Frequency
1.0　FLOORS	
1.1　**Carpet**	
full vacuum	daily
1.2　**Vinyl**	
dust control mop	daily
damp mop	daily
spray clean	weekly
1.3　**Granite**	
dust control mop	daily
spray clean	daily
1.4　**Entrance Mat**	
vacuum clean	daily
2.0　PAINTWORK, WALLS, PARTITIONS AND DOORS	
3.0　FURNITURE, FIXTURES AND FITTINGS	
4.0　REFUSE	

Figure 4.4 Typical cleaning schedule for paintwork, walls, partitions and doors in OFFICE areas

CLEANING SCHEDULE – OFFICES   CATEGORY CODE OF

Cleaning Requirements	Frequency
1.0 FLOORS	
2.0 PAINTWORK, WALLS, PARTITIONS, DOORS	
spot wipe doors	weekly
clean and polish dry glass doors	weekly
clean door pushplates, handles, kickplates	weekly
clean and polish dry glass in doors	weekly
spot wipe paintwork, walls, partitions and pillars	weekly
spot clean internal glass partitions and panels	weekly
spot clean upholstered screens	weekly
damp dust doors including frame and furniture	monthly
3.0 FURNITURE, FIXTURES AND FITTINGS	
4.0 REFUSE	

Figure 4.5 Typical cleaning schedule for furniture, fixtures and fittings in OFFICE areas

CLEANING SCHEDULE – OFFICES   CATEGORY CODE OF

Cleaning Requirements	Frequency
1.0 FLOORS	
2.0 PAINTWORK, WALLS, PARTITIONS, DOORS	
3.0 FURNITURE, FIXTURES AND FITTINGS	
clean mirrors	daily
dust hi-tech and presentation equipment	daily
damp wipe telephones	daily
clean and polish dry glass topped tables	daily
damp wipe external surface of vending/drinks machines, water dispensers	daily
damp dust tops of all furniture, window sills	daily
damp dust framework of furniture, fixtures and fittings, polish dry any metalwork	weekly
damp dust ledges, shelves, skirtings, fire equipment casing, radiators, wall fixtures and fittings and vents below 2m	weekly
clean and dry metal fittings	weekly
damp wipe and dry leather chairs	weekly
damp wipe free standing fans and lights	weekly
remove debris from upholstered furniture and spot clean if required	weekly
clean and polish dry furniture glass	weekly
damp wipe and dry leather topped desks and tables	weekly
4.0 REFUSE	

Figure 4.6 Typical cleaning schedule for refuse in OFFICE areas

CLEANING SCHEDULE – OFFICES   CATEGORY CODE OF

Cleaning Requirements	Frequency
1.0 FLOORS	
2.0 PAINTWORK, WALLS, PARTITIONS AND DOORS	
3.0 FURNITURE, FIXTURES AND FITTINGS	
4.0 REFUSE	
empty and damp wipe ashtrays	daily
empty paper collection wastebins and dispose of recyclable paper to designated area	daily
empty wastebins and dispose of rubbish, replace liner if required (at least weekly)	daily
damp wipe external surface of swing bins	daily
damp wipe wastebins	monthly

Figure 4.7 Typical cleaning schedule for sanitary fittings in TOILETS

CLEANING SCHEDULE – TOILETS   CATEGORY CODE TO

Cleaning Requirements	Frequency
1.0 FLOORS	
2.0 PAINTWORK, WALLS, DOORS, GLASS	
3.0 FURNITURE, FIXTURES AND FITTINGS	
4.0 SANITARY FITTINGS	
clean wash-basins, toilets, urinals, showers, drinking fountains and splashbacks, polish dry metal fittings	daily
damp dust toilet roll holders, soap dispensers, sanitary disposal and dispenser units, hand driers, roller towels and paper towel holders	daily
replenish consumables	daily
5.0 REFUSE	

Figure 4.8 Standard times used by Roger Blackburn for calculating the cost of different cleaning options at Neptune Stereophonics

Operation	Time allowed per 100 square metres (hours)
Spot vacuum	0.06
Full vacuum	0.14
Vacuum clean edges and corners	0.03
Damp dust tops of desks, tables and cupboards	0.03
Damp dust framework (detail dust)	0.20
Spot wipe paintwork, walls	0.03
Empty wastebins and replace liner if required	0.04
Damp wipe wastebins	0.05

Figure 4.9 Blackburn's first option for carpeted offices

Operation	Time for 100 sq.m. (hours) per operation		per annum	Time for 785.22 sq.m. per annum (hours)	
Daily operations					
spot vacuum floor	0.06	*(x 260 =)*	15.6	*(x 7.8522 =)*[1]	122.49
dust tops of desks, tables and cupboards	0.03	*(x 260 =)*	7.8	*(x 7.8522 =)*	61.25
empty wastebins and replace liner	0.04	*(x 260 =)*	10.4	*(x 7.8522 =)*	81.67
Weekly Operations					
full vacuum floor, vacuum edges and corners	0.14 +0.03 ──── 0.17	*(x52 =)*	8.8	*(x7.8522 =)*	69.41
spot wipe paintwork, walls	0.03	*(x52 =)*	1.6	*(x7.8522 =)*	12.25
detail dust	0.20	*(x52 =)*	10.4	*(x 7.8522 =)*	81.66
damp wipe wastebins	0.05	*(x52 =)*	2.6	*(x7.8522 =)*	20.42
TOTAL HOURS REQUIRED PER ANNUM					449.15

COST OF SERVICE = £449.15 x R
(where £R is the hourly rate of pay)

[1] Since the times are given per 100 square metres the multiplication factor is 785.22/100 or 7.8522 which is the number of *hundreds* of square metres of carpeted offices.

Figure 4.10 Blackburn's second option for carpeted offices

Operation	Time for 100 sq.m. (hours) per operation	per annum	Time for 785.22 sq.m. per annum (hours)
Daily operations			
full vacuum	0.14	36.4	285.82
detail dust	0.20	52.0	408.31
empty wastebins and replace liner	0.04	10.4	81.67
Weekly Operations			
vacuum edges and corners	0.03	1.6	12.25
spot wipe paintwork, walls	0.03	1.6	12.25
damp wipe wastebins	0.05	2.6	20.42
TOTAL HOURS REQUIRED PER ANNUM			820.72

COST OF SERVICE = £820.72 x R
(where £R is the hourly rate of pay)

4.11 Typical Day Cleaning Schedule for Neptune Stereophonics

DAY CLEANING SCHEDULE

Two day janitors are required to be on site at all times between the hours of 0800 and 2200, seven days per week. They will have the following responsibilities:

Toilets

Visit every toilet at least once during each production shift to undertake a check clean as follows:

> replenish consumables
> wipe clean wash–basins, splashbacks and surrounds
> wipe clean WCs and urinals
> spot wipe or mop as appropriate, any stains and spillages
> empty wastebins and replace liner (if necessary)

Restaurant

Visit the staff restaurant between the hours of 1400 and 1500 to spot mop floor, spot clean spillages on carpet or upholstery, to vacuum carpet and empty wastebins.

Respond to requests for assistance from catering management staff outside these times as necessary.

Locked Rooms

Clean any rooms that are inaccessible during the normal cleaning shift. These rooms may have been locked for security purposes and may require the presence of security personnel during the cleaning operation.

Emergencies

Carry out any emergency cleaning operations as the need arises. This may include damp mopping of spillages and the removal of stains from furniture or carpets as they occur and as notified by Neptune Stereophonic's staff.

General Duties

Assist in general duties as specified by the Facilities Manager or his designated representative.

5

PREPARING THE CLEANING SCHEDULES – II SPECIAL OPERATIONS AND PERIODICS

Whereas Chapter 4 was concerned with the development of schedules for core cleaning operations with a frequency of not less than monthly, in Chapter 5 we shall consider the scheduling of periodics and certain special operations which justify a separate specification in their own right.

SPECIAL OPERATIONS

What are special operations?

They are those tasks which, because of their importance, or magnitude, or cost, justify more precise scheduling or definition, than might normally apply if included in the core or periodic schedules.

They may also be operations that could well be the subject of separate contracts between the client and different contractors; or may be those that the contractor himself may decide to subcontract to specialists. Examples are:

- Carpet cleaning
- 'Hi-Tech' cleaning
- Glass Cleaning
- Cleaning external areas
- Deep cleaning in kitchens
- Deep cleaning in toilets
- Waste management
- Grounds maintenance

Some question whether such operations should form part of the principal cleaning specification at all and thus whether they should be negotiated during the main tendering process. There are a number of reasons why they should.

1. Pulling all of the special requirements into one contract makes budgeting easier.

2. Having to liaise with only one contractor reduces the amount of time spent managing contracted out cleaning operations as compared with the time that would be spent if they were the subject of separate contracts.

3. The 'buck stops' with the main contractor. Poor performance by sub-contractors becomes his problem rather than yours.

4. Less time, and hence money, is spent on the tendering procedure.

Let us consider each of the above operations in some detail.

Carpet Cleaning

As we have seen, vacuum cleaning and the removal of stains and spillages from carpets, form part of the core cleaning schedule. If the carpets are few in number, variety or usage then any periodic cleaning that is required may simply be included in the general *Periodic Schedule* as will be discussed later in this chapter. Such is the case at Neptune Stereophonics. Here, all of the carpets are supplied by the same manufacturer. They are similarly dark in shade and, with the exception of Reception which needs to be maintained to a high standard, and the Conference Room which has only infrequent usage, they all require the same degree of maintenance. Thus, it would be adequate for Roger Blackburn to schedule all of the carpet cleaning as part of his periodics rather than as a separate special operation. He is in fact unable to do this as we shall see shortly.

However, if the carpets vary significantly in colour from area to area, differ markedly in the amount of traffic to which they are subjected, or require maintenance to widely different standards then it may be desirable to schedule carpet cleaning separately.

By doing so you will also have the opportunity to examine in some detail the total cost of maintaining your carpets, thereby allowing you to manipulate frequencies in order to trim costs if the total contract price is ultimately too high. Indeed this benefit applies to all of the separately scheduled special operations, as it does to a lesser extent to all the periodics as we shall see later.

There is another reason why you may have to provide a separate schedule; the reason why Roger Blackburn cannot include the carpets in his periodics.

In the UK, at least one prominent manufacturer of contract carpets stipulates precisely how his carpets should be maintained. Furthermore, failure to maintain them in this way makes his guarantee invalid. In such cases it is imperative that the manufacturer's recommendations are followed. If the manufacturer also requires that you use a nominated carpet cleaning contractor you will have to ensure that your main contractor subcontracts the carpet cleaning to that named company. Otherwise you should speak directly to the carpet manufacturer and try to obtain his approval for *your* choice of contractor. Or buy someone else's carpets.

Figure 5.1 shows a typical carpet cleaning schedule. You will note that, like the schedules for core cleaning illustrated in the previous chapter, it provides information relating to both task and frequency. However, you will also note that it provides additional information concerning the frequencies for cleaning different categories of room depending upon their usage or importance.

Certain new terms have been introduced:–

Interim Maintenance is intended only to improve the visual appearance of the carpet, i.e. to remove superficial soil. It need only be undertaken on carpets likely to change shade if they become soiled. For example, pale carpets will become darker as they get dirty, and dark carpets may become lighter as they get dirty. In either case interim maintenance can be used as a cheaper option to extend the time period between the more costly deep cleaning techniques.

There are two interim techniques, bonnet buffing and absorbent powder cleaning. The former uses a rotary machine and soft pad moistened with a detergent solution. The machine is passed swiftly over the surface of the carpet with the objective of transferring soil from the pile to the pad. Powder cleaning involves brushing a powder into the carpet and then vacuuming it out again. The powder is impregnated with a mixture of solvent, water and detergent which loosens soils from the surface of the fibre. The soil is then absorbed by the powder and thus removed during the vacuuming operation.

If you have a preference for one technique or the other, specify it in your carpet cleaning schedule. Otherwise leave the choice to your cleaning contractor who will select the technique he best likes to use. Remember however that neither system can cope with heavy soiling conditions and occasional wet cleaning will be necessary.

There are two wet cleaning techniques, shampooing and spray extraction. Somewhat confusingly these are commonly known as periodic techniques. In the former, shampoo is brushed into the carpet using a rotary or cylindrical brush machine, is allowed to dry, and is then vacuumed away. Spray extraction involves spraying detergent into the carpet and almost simultaneously extracting it again. Spray extraction removes more soil but is more expensive than shampooing because it is slower. However, today it is the most common wet cleaning technique. Again, if you have a preference for one technique or the other, specify it in your carpet cleaning schedule.

Finally, when scheduling your carpet cleaning avoid the temptation to overclean. There is no need to clean carpets that do not show soiling or are perhaps positioned some considerable distance from external doors, more often than is necessary.

Hi-Tech Cleaning

Figure 5.2 is an illustration of the Hi-Tech cleaning schedule for Neptune Stereophonics. As you will see no frequency is less than monthly and you could thus easily incorporate these tasks into your core cleaning operations. If you have few items of Hi-Tech equipment then you will almost certainly do so.

– *sanitise telephone handset*

for example, might well appear as a task in the core schedules drafted as described in Chapter 4.

In some organisations staff are required to clean their own VDU screens and word processor keyboards and are issued with a 'kit' for this purpose. In practice they seldom make the effort and it is therefore far more desirable for cleaning to be the responsibility of the contractor.

By drafting a separate schedule for Hi-Tech equipment there are a number of advantages:-

1. VDU screens and keyboards are less likely to be neglected.

2. With a separate schedule it is easier to cost the operation (and thus manipulate frequencies if the total contract price is too high).

3. With a separate schedule, the contractor is more likely to designate one member of staff to be responsible for undertaking the work. This will inevitably bring greater expertise to the operation; and lessen the risk of inappropriate, or even damaging cleaning materials being used.

4. Monitoring to ensure that the work is actually carried out, tends to be easier.

Glass Cleaning

It is commonplace to find that window cleaning has been negotiated as a separate contract. Yet all of the arguments enumerated above apply equally well to window cleaning as to any other operation that can be subcontracted. Furthermore, if window cleaning forms part of the total specification there is no need to be concerned about where one contractor's responsibility ends and the other begins.

So you do not need to worry who should clean glass doors or internal glass partitions. It becomes the main contractor's problem. In other words, your only concern is to ensure that the job has been done, not who has done it.

Many of the larger contractors have their own window cleaning division. They are likely to visit the site during the preliminary stages of tendering and will decide whether your cradle is adequate or if they will need a 'cherry picker', 'magic carpet', bosun's chairs, ladder or abseiling gymnast. Smaller contractors will, if the job is in any way out of the ordinary, ask a specialist window cleaning contractor to accompany them at the time of the initial site visit. Whichever applies, the drafting of a glass cleaning schedule is desirable. These are not particularly complex as Figure 5.3 shows.

External Areas

The cleaning of external areas is often overlooked. Loading bays, car parks, paths and walkways, steps and external signage may all be forgotten. Yet some of these areas are the ones that visitors first encounter. Litter, bird droppings, cigarette ends (especially when tipped out of car ashtrays) present a most unsatisfactory first impression. The schedule for external cleaning shown in Figure 5.4 is an example of what might be drafted.

Deep Cleaning of Kitchens

Although you will ask for a separate quote for the periodic deep cleaning of the kitchen (see Chapter 9) it is unlikely that you will need to schedule individual operations in the way described above. Normally sufficient information can be given in the *General Notes* section of the specification for the contractor to be clear of your precise expectations. (See Chapter 6). However if there is any possibility of confusion, perhaps because there are several food preparation areas on site, or because the equipment is varied and complex, it may be appropriate to draft a separate schedule in order to make your requirements perfectly clear.

Deep Cleaning of Toilets

A similar argument applies to the deep cleaning of sanitary fittings, the requirements for which can also be specified in the *General Notes* unless the arrangements are particularly unusual – perhaps because there are swimming pool filters and traps that need special attention for example.

Waste Management

It is not within the scope of this book to discuss the principles of waste management in any great detail. Furthermore in most office blocks the types of waste that are generated are often simple and can be dealt with as part of the normal core cleaning schedule with some additional explanation in the *General Notes* as necessary.

Typical wastes include paper; shredded confidential waste, cardboard; aluminium cans, sanitary towels; general kitchen waste and scrap; most of which finishes up in the skip outside with only sanitary towels and in some cases kitchen waste, receiving any special treatment.

However, even in offices there is a growing trend towards recycling, if not for financial reasons then because there is an increased environmental awareness amongst staff. Thus some primary sorting is necessary, either by those producing waste, or by the contractor. Clearly the first option is the cheapest.

Where this trend for recycling exists, separate bins or carts may be positioned strategically around the site from whence the contents are taken by the contractor either for compaction and subsequent recycling, or for general disposal. If the sorting is good enough and the volumes of recyclable waste are high enough then a net contribution may arise from its sale. This can then be offset against the cost of handling and disposing of these wastes which are of no intrinsic value. (In some establishments, income from recyclable waste is given to charity).

The more involved your procedures for waste handling become then the more detailed a specification the contractor will require to enable him to deal with the waste according to your demands. If you consider therefore, that explicit instructions are necessary, draft a separate schedule for the waste handling.

Whatever your preference, you will need to meet the requirements of the Environmental Protection Act 1990 or any subsequent amendments and observe the Duty of Care Regulations 1991. In this context your waste is 'controlled' waste and must therefore be documented on transfer forms that log its history as the popular cliche has it, 'from cradle to grave'.

If your company generates quantities of waste that must be disposed of in a particular way in order to satisfy the requirements of the Environmental Protection Act then for sure you will need a separate schedule for waste management. Take a look at Figure 5.5 which is part of the schedule for a manufacturer of household detergents and cleaning products. The complete schedule might specify the disposal of metal drums, plastic drums, clean cardboard and paper, cardboard contaminated with chemical residues, faulty product, inflammable product in damaged aerosols, biohazard including 'sharps', and polythene, together with any number of other named wastes.

When wastes need such specialised handling it may well be argued that they should not be included in the cleaning specification at all but should form part of a quite separate waste management contract tendered only by specialist waste management contractors. On the other hand, if the principle of reducing main contractors to a minimal number is accepted as has been proposed earlier in this chapter, then some cleaning contractors may find themselves quoting against specifications that do indeed include specialist waste disposal procedures.

Grounds Maintenance

It is not difficult to absorb grounds maintenance into the cleaning specification thereby delegating the responsibility for management to the cleaning contractor. He will almost certainly subcontract the work unless it is quite straightforward. The full schedule will of course detail activities throughout the entire year, from weeding and feeding in the spring to leaf sweeping in late autumn. Anyone with a good book written for the amateur domestic gardener should be capable of drafting a suitable document and it is not intended to reproduce that type of information in the present text. Bear in mind however that some areas of your site may require different frequencies of ground maintenance than others. Lawns by reception may need mowing on a weekly basis whilst outlying grassy areas may be coarse cut only every four weeks for example. A colour coded plan of the site may usefully designate areas that require fundamentally different frequencies or techniques.

PERIODIC OPERATIONS

Periodic operations are normally carried out with a frequency of less than monthly and should be costed separately task by task during the tendering process. Part of the *Periodic Schedule* for Neptune Stereophonics is shown in Figure 5.6. You will see that frequencies ranging from 3 monthly (i.e. 4 operations per annum), to annually have been included and in general terms the procedures are concerned with floors, vertical surfaces, fixtures and fittings, and ceilings.

You should have taken sufficiently adequate notes during your building survey to enable you to draft a schedule for periodic operations without moving from your desk. Choosing frequencies may require a little more careful thought however since periodic cleaning may represent a significant component of the total cost. Fortunately over-specifying the frequencies is not too catastrophic for there is always an opportunity to reduce them once you have obtained a price per operation during the tendering stage.

Take care not to leave anything out, only to find halfway through the contract that you have omitted something vital. The cost of adding it later, once you are through the competitive tendering procedure, can prove to be more costly than you bargained. To minimise this risk, it is a good discipline to work through your possible requirements in the same way that you drafted the schedules for the core cleaning operations – starting with the floor and working up. Some options are considered below.

Floors

If the floors are carpeted:

You may already have drafted a separate schedule for carpet cleaning as a special operation, as has been discussed earlier in this chapter. If this is the case then carpets need not be included on the *Periodic Schedule*. If you have not then your interim (if any) or wet cleaning requirements should now be included. Choose from:

- *bonnet buff carpets*
- *powder clean carpets*
- *spray extract carpets*
- *shampoo carpets*
- *clean oriental rugs*

If the floors are smooth:

A number of operations may apply, depending upon whether the floors are sealed, polished, contaminated with thick greases (as in a garage); wood, terrazzo, vinyl, marble etc. Any of the following may be required:–

- *neutral scrub*[1]
- *neutral scrub and polish*
- *strip and redress*
- *sand and reseal*[2]
- *vitrify*[3]
- *scarify*[4]
- *sweep*

Vertical Surfaces

Paintwork, walls, doors and glass, and items associated with these. Include partitions, goods lift sides, upholstered screens, photocell activated factory doors, fire shutter guides and any number of items that fall loosely into this category. Also relevant is that items over 2 metres high that require cleaning usually appear on the *Periodic Schedule*. Tasks you may want to specify include:

- *vacuum clean fabric wall coverings*
- *damp wipe and dry partitioning between offices*
- *spot clean and vacuum clean upholstered screens*
- *spray extract upholstered screens*

---

[1] In busy areas this may be sufficiently frequent to be a core cleaning operation
[2] Wood floors   [3] Marble, terrazzo   [4] Concrete

- *hose brush factory doors*
- *wash and dry goods lift sides*
- *wash walls from ceiling to floor*
- *damp wipe and dry marble walls from ceiling to floor*

Furniture, Fixtures and Fittings

Typical operations are:

- *vacuum clean upholstered chairs*
- *scrub vinyl chairs*
- *spray extract upholstered furniture*
- *remove, dry clean and re-hang curtains*
- *damp wipe venetian blinds*
- *remove venetian blinds for ultrasonic cleaning*
- *vacuum clean vertical blinds*
- *vacuum clean roller blinds*
- *high dust all fixtures and fittings above 2m*
- *high dust high racking[1]*
- *damp wipe all pipework and machinery housings (except electrical switchgear)[2]*
- *damp wipe external surfaces of Eurocarts*
- *clean external surfaces and glass of fume cupboards*

Ceilings

Include:

- *dust suspended spot/strip lighting*
- *clean diffusers and change light tubes[3]*
- *vacuum clean overhead pipework and ducting*
- *vacuum clean ventilation grilles*
- *remove and clean ventilation grilles*
- *vacuum clean or wash ceiling tiles according to type*
- *remove algae from external flat roofs*

Clearly, in each of these categories, a large number of options are available. Most will suggest themselves to you during your building survey. Do not forget that your primary objective is to specify *every* cleaning operation that you require.

---

[1] In a warehouse for example.
[2] In plant rooms for example.
[3] If you change lighting tubes as part of a preventive maintenance programme, it makes sense to do it when the diffusers are cleaned.

SUMMARY

Operations which are carried out less frequently than monthly or which are considered important enough to warrant scheduling separately, may be classified as periodics or special operations.

Typical special operations include carpet cleaning, hi-tech cleaning, glass cleaning, cleaning external areas, deep cleaning in kitchens and toilets, waste management and grounds maintenance.

By treating special operations and periodics separately from the core operations there is an opportunity to cost each task individually. This gives you additional flexibility when trying to tailor your cleaning requirements to your budget.

Drafting a periodic schedule should be a systematic procedure making use of the notes you made at the time the building was surveyed in order to complete your Schedule of Accommodation and to schedule the core cleaning.

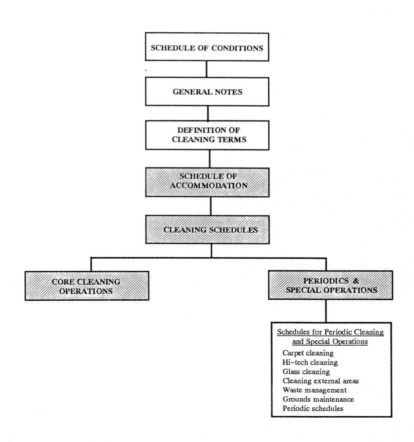

Figure 5.1 Example of Schedule for Carpet Cleaning

Area	Frequency of interim maintenance[1]	Frequency of periodic maintenance[2]
RECEPTION	Fortnightly[3]	3 monthly
LIFT LOBBIES	Monthly	3 monthly
LIFTS		3 monthly[4]
RESTAURANT	Monthly	3 monthly
CIRCULATION	2 monthly	6 monthly
OFFICES		annually[5]

1. or 'Bonnet Buffing' or 'Powder Cleaning'
2. or 'Shampooing' or 'Spray Extraction'
3. assuming carpets are fairly light in shade
4. or put entrance mats in the lifts and change them weekly (daily if you incorporate a 'have a nice Tuesday etc.' message)
5. assumes dark shades

Table 5.2   Example of Schedule for Hi-Tech Cleaning at Neptune Stereophonics

TELEPHONES/PUBLIC TELEPHONES

Wipe body and handset of telephone with a mild disinfectant solution or telephone sanitising wipes. Special care is required to clean the cord around the keys. Damp wipe coin box and hood, if applicable. Damp wipe fixtures and fittings of public telephone stand, polish dry any metalwork, if applicable.

FREQUENCY: PUBLIC TELEPHONES   daily
              TELEPHONES   fortnightly

FACSIMILE MACHINES

Wipe machine casing and telephone handset (if applicable), paper trays etc. with an approved spray cleaning/sanitising agent and lint free cloth. Damp wipe and dry perspex cover (if applicable) using a cloth moistened with neutral detergent solution. Polish dry to remove cleaning marks.

FREQUENCY: monthly

TERMINALS/VDUs/MICROFICHES

Ensure that the screen is turned off before cleaning. Clean the screen with a soft cloth and proprietary antistatic glass cleaner, followed by a cloth moistened with water. Wipe the screen dry with a clean lint-free soft cloth. Wipe down the case of the terminal with an approved spray cleaning agent. Damp wipe the keyboard and connecting lead.

FREQUENCY: monthly

T.V. SCREENS

Ensure that the screen is turned off before cleaning. Clean the screen with a soft cloth and proprietary antistatic glass cleaner, followed by a cloth moistened with water. Wipe the screen casing with an approved spray cleaning agent.

FREQUENCY: monthly

PRINTERS

Wipe printers and printer hoods using an approved spray cleaning agent and lint-free cloth.

FREQUENCY: monthly

PHOTOCOPIERS/SHREDDERS

Wipe machine casings/paper trays etc. with an approved spray cleaning agent and lint-free cloth.

FREQUENCY: monthly

AUDIO VISUAL MIRROR

Clean with the manufacturer's approved cleaning agent using a lint-free cloth and in accordance with the manufacturer's recommendations.

FREQUENCY: monthly

Table 5.3   Example of Schedule for Glass Cleaning

GLASS CLEANING SCHEDULE

On completion of the following operations all adjacent surfaces e.g. sills, mullions and frames, should be wiped clean and left free from smears. The contractor should allow for the removal and repositioning of furniture as necessary to allow safe access to windows.

INTERNAL GLASS

2 monthly

Clean both sides of all internal glass partitions and panels.

Clean internal surfaces of all windows (three sides to be cleaned)* including skylights.

Damp wipe internal window sills between secondary glazing units.

EXTERNAL GLASS

2 monthly

Clean external surfaces of all windows.

Clean external surfaces of glass in all roofs where applicable.

---

*   This example presumes all windows have secondary glazing. There are thus three sides to be cleaned internally.

Table 5.4   Example of Schedule for External Cleaning

CLEANING SCHEDULE – EXTERNAL AREAS

Daily

> pick up litter around perimeter of buildings
> sweep entrance steps
> sweep loading bay
> remove graffiti
> clean and polish brass 'registered office' plaque
> damp dust front door, clean and polish dry all bronzework

Weekly

> sweep courtyard, car park, ramps and all metal staircases
> remove leaves from alcoves (as necessary)
> clean birdlime from paviours at entrance
> empty and damp wipe external bins and replace liner

Monthly

> wash down railings and wrought iron gate
> hose down metal staircase
> hose down compactor yard
> clean external signage

Table 5.5 Example of part of a Waste Management Schedule for a manufacturer of household detergents

WASTE MANAGEMENT SCHEDULE

200l plastic or metal drums

Drums containing flammable materials

1. As necessary, arrange uplift from racking and transport off-site to authorised incineration plant.

Drums containing non-flammable materials

1. As necessary arrange uplift from racking and transport off-site to designated landfill location.

Empty plastic or metal drums

1. Rinse out, discharging washings to effluent treatment system.
2. As necessary arrange uplift and transport off-site for subsequent re-use.

Polythene waste

Clear polythene waste

1. Supply and install 14 No. 1100 litre colour coded yellow wheeled carts in production and manufacturing areas.
2. Exchange cart when full for a replacement empty cart and transfer full cart to waste collection area.
3. Compact and bale polythene waste, first sorting any accidental contamination.
4. As necessary arrange uplift and transport off-site for recycling.
5. Wash Eurocart once each month as necessary.

Scrap metal waste

1. Supply and install 2 No. 15 cu.m. skips in workshop areas.
2. Empty when full
3. As necessary, arrange uplift and transport off-site for recycling or authorised landfill location as appropriate.

Test tubes and swabs

1. Uplift special laboratory bins
2. As necessary, arrange uplift off-site to appropriate incineration plant.
3. Replace with new bins.

Figure 5.6 Part of the Periodic Schedule for Neptune Stereophonics

PERIODIC SCHEDULE

Task	Room Category	Frequency
Sweep Floors	Plant Rooms	3 monthly
Neutral scrub vinyl floors	All areas except production	3 monthly
Damp wipe partitioning between offices	Offices	3 monthly
Wash and dry goods lift walls	Circulation	3 monthly
Spot clean and vacuum clean upholstered screens	Restaurant Offices	2 monthly
Vacuum clean upholstered chairs	Reception Offices Production areas	2 monthly 3 monthly 6 monthly
Hose brush factory doors	Production and Warehouse	6 monthly
High dust high racking	Warehouse	6 monthly
Remove, dry clean and rehang curtains	Offices	Annually
Vacuum vertical blinds	Reception Offices	3 monthly 6 monthly
Damp wipe venetian blinds	Restaurant	6 monthly
Clean diffusers and change light tubes	All areas	Annually
Remove algae	External flat roofs	Annually
Clean overhead pipework and ducting	Production areas	Annually

6

WRITING THE GENERAL NOTES

So far we have discussed two essential components of the cleaning specification – the *Schedule of Accommodation* which details all the rooms to be cleaned, specifies the flooring type and quantifies the floor area; and the *Cleaning Schedules* which detail core cleaning, periodic cleaning and special operations. There is one further component that the contractor needs before he is able to effectively meet your cleaning requirements. This component is the **General Notes** section of the Specification.

The *General Notes* are intended to clarify any ambiguities and to offer more comprehensive information relating to any special aspects which may affect the contractor's ability to provide the necessary cleaning service or to quote.

In this chapter we shall consider the type of information which you will need to provide in the way of *General Notes* and discuss the rationale behind it. The explanation is extensively illustrated by reference to the Notes prepared by Roger Blackburn for Neptune Stereophonics which appear throughout the chapter. Each of the sub-headings which follow suggest the sub-heading you may wish to use when drafting your own *General Notes*. The text which appears in italics is Roger Blackburn's attempt.

Before you begin however, it is useful to remind yourself of the questions you answered in Chapter 3. For convenience these are reproduced in Figure 6.1.

Begin by describing the areas to be cleaned.

Areas to be cleaned

It may be that you do not want all of your building to be cleaned. You may have tenants occupying one floor, or part of a floor for example and your commitment as landlord may be to clean only their common areas i.e. stairs, lift lobbies and toilets.

There may be areas where the contractor is not allowed access: computer rooms, operating theatres, or security sensitive areas for instance.

This section in the *General Notes* clarifies where you require a cleaning service by directing the contractor to read the *Schedule of Accommodation*.

It is not essential to refer in this section to areas that do not need cleaning, since their omission from the *Schedule of Accommodation* infers that they are not to be included. Thus it is implicit at Neptune Stereophonics that since the kitchens do not appear on the *Schedule of Accommodation*, then they are to be excluded from the costings. Nevertheless it is as well to draw attention to this fact in this section to lessen the risk that you will be charged for tasks you do not require.

Roger Blackburn wrote:

Areas to be cleaned

All areas to be cleaned are detailed on the Schedule of Accommodation. The Schedule indicates the size of each room in square metres, the floor type, the category into which the rooms have been classified, and any special requirements. A plan of the site is also provided. This plan has been colour coded to assist in the interpretation of the document. It will be noted that the kitchen, dishwash area and a number of other small rooms do not require any cleaning service. These areas have been left uncoloured on the plans.

Cleaning Requirements

This section provides specific explanatory detail in relation to the following issues:–

- Times of cleaning, the number of days of service, and bank holiday requirements
- The classification of rooms into various categories
- Periodic cleaning
- Special operations
- Cleaning computer rooms

Let us consider these aspects in a little more detail.

Times of cleaning etc.

You must decide what times of the day you want the building to be cleaned and how many days per week the cleaning is required. There are a number of factors to be considered.

i   Do you want the building to be cleaned when staff are present or absent?

      Most favour the latter. There is less disruption to your own workforce. Dust will settle in an empty building rather than remaining airborne in traffic created turbulence. The cleaning staff are less likely to be interrupted and can therefore perform more effectively, e.g. not needing to turn off a vacuum cleaner because the telephone rings. There will be no gossiping between cleaning staff and building occupants. Cleaners are able to move around the building in a methodical fashion.

There are however some who favour having the cleaners present during the day. They argue that if cleaners can get to know the individuals they serve, they will be more highly motivated and will perform more effectively. It is also claimed that the cleaners identify with the client's corporate image rather than their employer's, that it is easier to guarantee performance if they are highly visible and that they can be redeployed more effectively to meet short term changing circumstances (e.g. an occupant may be ill or on leave and therefore does not require a cleaning service in his office for a number of days).

ii If the cleaning is to be undertaken outside normal working hours then do you prefer an early morning shift, or an evening shift – or even a night shift? Remember if you require cleaners between 0300 and 0700 they will be more expensive to employ, turnover and absenteeism may be high and they will be more difficult to replace. Furthermore shift patterns may be affected by the accessibility of the site; the convenience of public transport for example. Most cleaners do not have their own transport and therefore rely upon the availability of buses. Abnormal shift times may thus have some bearing on labour flexibility.

iii Shift length is also important. Generally, the longer the shift the greater the commitment to the job. Cleaners working on an eight hour shift are more likely to consider that they have a 'proper' job than those working fewer hours and significant reductions in absenteeism have been noted amongst the same group of staff when shift lengths have been renegotiated to occupy longer periods. National Insurance requirements may however, increase the overheads if longer shifts are worked. (See also Chapter 11 for more discussion about shift lengths).

iv Less difficult to decide is how many days per week you need the contractor to be on site. If you work a standard Monday to Friday week then clearly a five day service will be required from your contractor. If you work six days, such as may be the case in a department store for example, then he will need to be on site for six days each week. Greater difficulty may be encountered if you are a College which has a variable cleaning requirement throughout the year; or as in the case of Neptune Stereophonics, your working patterns are changing as your business grows.

v Finally you will need to advise the contractor whether a bank holiday service is required and if so, for how many days per annum.

With all these considerations in mind, Roger Blackburn began the section specifying his cleaning requirements as follows:–

Cleaning Requirements

A seven day per week service is required in production areas, circulation areas, toilets, the restaurant and the staff entrance. All other areas require a five day per week service. Core cleaning operations are to be carried out between 2300 and 0700. No service is required on bank holidays.

Blackburn has clearly made up his mind what time he wants the cleaners on site since his indecision in Chapter 4. He has decided to have them overlap the production shift change.

Room classification system

Next you should explain that the core cleaning operations have been scheduled according to a classification system. Blackburn continued:

Each room or area requiring a cleaning service has been classified according to its usage into one of several categories. These categories are listed in the Schedule of Accommodation. Rooms or areas listed under the same category as one another are each subject to the same cleaning schedule, details of which may be found in the appropriate section of this document. In the Schedules monthly tasks are to be carried out every calendar month i.e. twelve times per annum.

Schedules and their interpretation

As you will see from Blackburn's entry above, this paragraph has also introduced the principle of separate cleaning schedules for each category of room. Additionally he has clarified the term *monthly* to indicate that the operation should be carried out twelve times annually rather than every four weeks i.e. thirteen times per annum. Be clear what you want and specify it. If you really want *monthly* tasks to be carried out thirteen times per annum make sure that they are. Otherwise you may only receive 12/13ths of what you are paying for.

Whilst still concerned with clarifying aspects of the core cleaning operations it is useful at this point in the *General Notes* to introduce some 'catch–all' phrase to ensure that even if you have failed to itemise all of the furniture, fixtures and fittings in the *Cleaning Schedules* the contractor should not fail to clean them. A suitable way in which this may be done is as follows:–

The schedules do not identify specific types of furniture, fixtures and fittings to be cleaned. These may include any of the following:–

Upholstered furniture, leather furniture, antique wood furniture, counters, lami– nated desks, glass topped tables, coffee tables, office tables, display boards, wipe boards, upholstered screens, filing cabinets, cupboards, coatstands, light switches, plug sockets, free standing lights, fire equipment, signage, glass fronted cabinets, coat racks, lockers, handrails, balustrades, shelves, coat hooks, pigeon holes, free standing fans, desk lights, cloths, leaflet holders, air conditioning units, free standing signs, presentation equipment such as flip chart stands, overhead projec– tors etc.

Of course if you operate a successful 'clear desk' policy you may not need to include this instruction – assuming of course that all members of staff follow this house rule!

Periodic Cleaning and Special Operations

You are now ready to provide more specific detail in relation to the periodics or special operations that may be scheduled. It may be appropriate to enumerate them as sub-headings. For example:−

1. *Periodic Cleaning*

Periodic cleaning operations (i.e. operations undertaken with a frequency of less than monthly) are specified in a separate schedule which may be found on page p of the Specification and should normally be undertaken between the hours of 2300 and 0700. Each of the periodic operations is to be costed separately and these separate costings must be shown on page p of the Tender Return.

The contractor must identify how periodic work will be phased and must show his deployment of labour on the appropriate forms in the Tender Return. Periodic tasks are to be invoiced only upon completion.

There are a number of important points here.

i Blackburn has indicated that the periodic operations should be undertaken between the hours of 2300 and 0700.

Unless he has good reason for selecting these hours (perhaps because he feels they will cause less disruption amongst his own workforce) this is not an especially good time. Labour employed for periodics is not likely to be the same as that carrying out core cleaning operations unless the contractor intends to invite his staff to work overtime. If he asks them to work *overtime* then by definition they will have to work outside the hours of 2300−0700 thereby contravening the requirement laid down by Blackburn.

If the contractor employs separate staff to perform the periodic operations then wage rates for employees working between 2300 and 0700 are likely to be higher than for those working between 0900 and 1700. Blackburn should ask himself whether he really needs his periodic operations to be carried out in the middle of the night. Perhaps a compromise of 1700−2100 would be perfectly acceptable.

ii Each periodic operation is to be costed separately. As we have seen − and will consider again later − this gives Blackburn the opportunity to adjust his requirements to suit his budget whilst knowing at once what savings can be realised if he reduces his frequencies.

iii Blackburn has asked the contractor to show how he intends to deploy his labour for completing periodic operations. This subsequently will help Blackburn ensure that the contractor has given sufficient consideration to the scheduling and staffing for periodics.

iv 'Periodic tasks are to only be invoiced upon completion'. This provides a very important safeguard. Suppose for example the annual contract for the core cleaning operations is valued at £220,000 and carpet cleaning, to be carried out every three months, costs an additional £20,000 per annum. Suppose also for simplicity that there are no other periodic operations!

Then the total value of the contract is £240,000. If the contractor submits his invoice every calendar month based upon core cleaning and carpet cleaning then he will charge £20,000 per month. In the year he will receive £240,000.

If however he carries out only three carpet cleaning operations instead of the four that are scheduled but is paid only on completion of the periodics, then in the year he will receive only £235,000 i.e. £220,000 for core cleaning and £15,000 for carpet cleaning. Thus the contractor is paid only for the work he does rather than the work he is *supposed* to do.

Clearly, periodic operations should only be paid for on completion and against the signature of a duly authorised client representative.

v   After *should normally be undertaken between the hours of 2300 and 0700* in his explanatory notes, Blackburn should also have considered the possibility that even though these are his preferred times, there are some tasks that may require daylight or perhaps the total absence of staff in a particular area between these times. He could have added:

Where this is not possible it may be necessary to complete certain operations at different times or at weekends. In such cases permission must first be obtained from Neptune Stereophonics.

Day Cleaning Service

As we have already discussed certain operations need to be undertaken during the daytime (a typical schedule is illustrated by Figure 4.11). This is likely to be the case in all instances in any large building where cleaning is carried out outside the hours of 0900–1700. Some additional explanation should be provided at this point in the *General Notes*. A typical instruction, as drafted for Neptune Stereophonics, is as follows:-

2.   Day Cleaning Service

Rooms or areas which require a day cleaning service are shown on the Schedule of Accommodation under the column 'special requirements'. This service is required between the hours 0800 and 2200 seven days per week. These operatives shall perform the duties specified in the Day Cleaning Schedule which appears on page p of the Specification.

As well as carrying out the specified duties, day cleaners may be required to cope with any cleaning problems that may arise anywhere on the site. A flexible approach is required in order to perform this service successfully.

The weekend day cleaning requirement will be reviewed from time to time. Any significant reduction in weekend working by Neptune Stereophonics staff may result in a reduction in the weekend day cleaning requirement. A consequent reduction in contract price will be negotiated based upon rates quoted in the Tender Return.

Carpet Cleaning

Some of the problems associated with the scheduling of carpet cleaning have been discussed in Chapter 5 and in the specimen schedule shown in Figure 5.1 the footnotes serve to illustrate typical choices which have to be made.

It is not possible here to draft one comprehensive note that covers all contingencies. Nor is it possible to provide a specimen set of alternative notes that might apply. Instead you will need to think through your own requirements and clarify any ambiguities that might arise specific to your particular site. The following three specimens may help.

3. *Carpet Maintenance*

Spot and stain removal of all carpeted areas forms part of the core cleaning schedule. Day cleaning operatives may also be required to carry out this work. All operatives who may be involved in the removal of stains or spillages must therefore be trained in the necessary techniques. The contractor is expected to provide a suitable range of stain removal chemicals appropriate for use on the various types of carpet to be found within the building. He is also expected to provide, and retain on site, a small spray extraction machine for stain removal purposes.

If you find it impossible to draft a carpet cleaning schedule because the range of carpets is so varied in type, colour and traffic, you may prefer to supplement the previous note with one along the following lines.

Because of the variety of carpets used throughout the site, set frequencies for interim or periodic maintenance have not been established. Instead carpet cleaning will be carried out at the discretion of Neptune Stereophonics. The contractor is therefore required to provide a price per 100 square metres for interim maintenance[1] and another for periodic maintenance[2] on page p of the Tender Return.

If you face Roger Blackburn's problem that the carpet manufacturer dictates how your carpets should be cleaned and possibly whom should clean them, your note may read.

Interim and Periodic carpet cleaning do not form part of the Specification but will be undertaken by a professional carpet cleaner under the direct supervision of Neptune Stereophonics.

---

1. You may instead want to say *for bonnet buffing* or *for powder cleaning*.
2. Or *spray extraction* or *shampooing* according to your preference.

Hi-Tech Cleaning

There are a number of points you will need to make:-

    i   That there is a separate specification for the Hi-Tech cleaning.

    ii   That a separate price is required

    iii   That trained personnel using prescribed chemicals are to be involved.

You will also need to show the approximate numbers of each item of equipment specified in the Hi-Tech schedule so that an accurate cost can be submitted.

Roger Blackburn's note for Neptune Stereophonics reads as follows:-

4.   *Hi-Tech Cleaning*

A separate schedule has been provided for Hi-Tech cleaning. This appears on page p of the Specification and includes telephones, facsimile machines, VDUs, printers and closed circuit television screens. A separate costing is required for this service and should be shown on page p of the Tender Return.

The cleaning of Hi-Tech equipment as shown in the relevant schedule must only be carried out by competent staff adequately trained for this work. Only cleaning materials specified as suitable for Hi-Tech equipment must be used. Telephones must be sanitised using a proprietary sanitiser. The cleaning agents and sanitisers that are to be used must be shown on the list of 'Materials to be Used' which appears on page p of the Tender Return.

The approximate numbers of each type of equipment are as follows:-

Public Telephones	4
Telephones	320
VDUs/terminals and keyboards	300
CCTV screens	10
Microfiches	4
Printers	180
Photocopiers/shredders	16

5.   Glass and Window Cleaning

Much of the internal glass (except at windows) will be cleaned as part of the core cleaning operation. Thus, glass partitions, glass doors, and vision panels in doors, will not have been separately scheduled.

The schedule shown in Figure 5.3 is quite straightforward and would only be complicated if for example, there were several different buildings or areas within the same building requiring different frequencies.

It may also be important to draw attention to Health and Safety aspects passing as much responsibility as you are legally able, onto the contractor:-

5. *Glass Cleaning*

The cleaning of all glass partitions and internal glass panels forms part of the core cleaning schedule and should be carried out between the designated hours of 2300 and 0700. Window cleaning may only be carried out during daylight hours however and forms the basis of a separate schedule shown on page p of the Specification. A separate price is also required, to be shown on page p of the Tender Return.

The contractor must remove and replace any items of furniture as necessary to gain access to the windows. Items on ledges and window sills must not be removed or moved by cleaning staff unless permission is given by Neptune Stereophonics.

On completion of window cleaning operations all adjacent surfaces, sills, mullions, frames etc., should be wiped down to remove splashes and smears. A satisfaction note must be signed by Neptune Stereophonic's appointed representative before any payment will be authorised.

The contractor must provide all necessary equipment to carry out window cleaning in a safe manner with adequate supervision and which meets prevailing Health and Safety legislation. If in doubt it is the contractor's responsibility to seek specialist advice from the Health and Safety Executive or other official body. Neptune Stereophonics may also make enquiries to or invite inspection by the appropriate authorities should they consider that the need arises.

You should also provide details of any access equipment already available and note any windows that cannot be accessed. Stipulate also who is responsible for maintenance of access equipment etc.

<u>External Areas</u>

Depending upon the times selected for cleaning it may be necessary to draw attention to the fact that external areas may need cleaning at some other time i.e. during daylight hours. You may also wish to specify a preference for a particular type of cleaning machine. If you have a car park the size of a football pitch for example you will not want it cleaning with a pedestrian operated sweeping machine. (You may be surprised to learn that some contractors do indeed specify such a machine for this type of operation). A simple note may be all that you need.

6. *Cleaning External Areas*

Cleaning of External Areas is detailed on page p of the Cleaning Specification. This work must be carried out during daylight hours. Car park sweeping will be carried out only at weekends.

The contractor is expected to provide, or hire suitable equipment for sweeping car parks and access roads and must specify which equipment he intends to use on page p of the Tender Return.

Deep Cleaning of Kitchens

If you have a separate catering contractor as is the case at Neptune Stereophonics it may be that he is responsible for the deep cleaning operations as well as day to day cleaning in food preparation areas.

Find out. You do not want to pay for the same operation to be carried out by two separate contractors. Furthermore the chances are one of them will decide not to do it but will try to charge you for it nevertheless.

Deep cleaning in kitchens is an important task. Not only are there hygiene connotations but failure to regularly clean ducts and grease filters may present a fire risk. A typical note may read:–

7. Deep Cleaning in Kitchens

A separate costing is required for the deep cleaning of the kitchen, servery and dishwash areas which is to be carried out four times per annum. The operation must include the cleaning of all walls and floors as well as catering equipment but not refrigerators or freezers). Catering equipment includes cookers, hotplates, ovens, steamers, salamanders, garbage disposal units and grease filters in extract systems. It is the contractor's responsibility to ensure that he identifies every item that needs to be cleaned.

In fact this is a rather simplistic approach to kitchen cleaning. In practice this note should be drafted after full consultation with your catering manager.

A similar comment may also apply tor deep cleaning in toilets.

The following note expresses a minimum requirement. (Water hardness is the primary determinant).

8. Deep Cleaning of Toilets

A separate costing is required for the deep cleaning of toilets including WCs, urinals, washbasins, showers, sink units and extractor grilles. Cleaning involves the removal of organic and inorganic deposits from those areas not normally accessible during regular cleaning. The service is required twice per annum, each of which must be costed separately and shown on page p of the Tender Return.

Computer Room Cleaning

It is quite likely that you will have classified computer rooms into a category of their own and will thus have drafted a separate schedule of core cleaning operations as part of the exercise described in Chapter 4. If this is the case, it probably means that your computer room is a sensitive area needing special instructions to ensure that no mistakes are made. In general terms you will probably require:–

    i   minimal use of water

ii	competent cleaning operatives

iii	anti-static materials and equipment

iv	special filtration on the vacuum cleaners

v	notification before cleaning is carried out especially in the case of periodic operations

vi	that power savers should not be disconnected (to plug in a vacuum cleaner for example)

Obviously you will need to speak to the relevant head of department to ensure that you make no errors of omission when drafting your rules.

The instructions Roger Blackburn drafted were as follows:

9. Computer Room Cleaning

The Computer Room has been classified as a separate category (referenced CO) and therefore has its own cleaning schedule which is shown on page p of the Specification.

This area is sensitive to dust and water and should only be cleaned by competent adequately trained staff. The contractor must strictly control cleaning methods and the materials which are used.

All core cleaning operations must be carried out within the normal shift period from 2300 – 0700. The Computer Room is in use during these hours and cleaning operatives must liaise with Neptune Stereophonics personnel using the room before any cleaning is commenced.

The following additional rules will apply:–

i	All cleaning materials and equipment must be anti-static.

ii	Vacuum cleaners must be fitted with triple filters such that no particle larger than 0.03 microns can be emitted to the atmosphere.

iii	No plug shall be removed from any socket. No cables shall be dis– connected.

iv	All cleaning to be undertaken must be approved by the IT Manager or his nominee notwithstanding any instruction given here or in the computer schedule shown on page p of this specification.

v	Cleaning operatives must take care not to disturb computer room staff whilst equipment is on–line. Such equipment will only be cleaned with the express permission of the computer operative.

> vi The Periodic Schedule specifies the vacuuming of floor voids. The Facilities Manager at Neptune Stereophonics or his nominated representative must be given 48 hours notice before this work is undertaken so that all smoke detectors and fire extinguishing systems in the void can be deactivated before vacuuming commences.

Remember, once you have drafted your notes on computer room cleaning, show them to the relevant head of department. A mistake could be too serious to contemplate.

Light Fixtures and Fittings

You may well have instructed the contractor to clean light diffusers and fittings as part of your periodic schedule. Some companies take this opportunity to require the contractor to replace light tubes as part of a preventive maintenance programme, i.e. rather than face the random disruption of replacing tubes as they fail, they have a formal policy of changing all tubes (and starters) when the light is being accessed for cleaning. In sites where access is difficult e.g. over production lines, this is thought to be the most cost effective policy.

10. Light Fixtures and Fittings/Tube Replacement

The cleaning of light fixtures and fittings forms part of the Periodic Schedule. The contractor will be required to change all fluorescent tubes and starters at the time this cleaning is carried out. Replacement tubes and starters will be supplied free of charge by Neptune Stereophonics.

If he had wished, Roger Blackburn could have asked his contractor to provide a cost for replacement tubes and starters. If he had done so, he would have needed to provide technical data and numbers i.e. the numbers of tubes of different lengths and their wattage.

If any further supporting information is required, perhaps in relation to periodics or special operations not discussed here, add them to this present section to conclude the 'Cleaning Requirements' component of the *General Notes*.

Grounds Maintenance and Waste Management, although discussed in Chapter 5 as part of the special operations should not appear as numbers 10 and 11 in the above list since they do not form part of the 'Cleaning Requirements' per se. Instead, discuss each one separately in its own right and under its own heading. To maintain continuity in the General Notes however, it is suggested that grounds maintenance is left until last, directly preceded by the notes relating to waste management. Before we consider either, there are two more aspects to be clarified:

Consumables

You will recall from Chapter 2 that at Neptune Stereophonics, Roger Blackburn had a procession of contractors attending to the toilets when in reality he required only one. In order to avoid this for the future he decided to delegate the responsibility to his cleaning contractor alone. To implement this you will note that in the day cleaning schedule illustrated in Figure 4.11 there is an instruction:–

Visit every toilet at least once during each production shift to undertake a check clean as follows:

– replenish consumables...

The same instruction appears in the *Cleaning Schedule* for toilets as shown in Figure 4.7.

Clearly, if you want the cleaning contractor to replenish your supplies, then you must either provide him with the supplies or require him to provide them. If you require him to provide them you will need to indicate:–

    i   What consumables are required

    ii   What quantity is required

The reasons for stipulating precisely what consumables are required are two-fold. One is to ensure that at the competitive tendering phase you will not be tempted to select a contractor who quotes say £2,000 per annum for toilet rolls in favour of one who quotes £4,000 only to find that the cheaper quote results in the supply of toilet rolls that are unacceptably inferior. The other reason is that you may have found your own preferences for consumables by trial and error over preceding months.

The reasons for indicating quantity are obvious. The contractor will need to know your typical utilisation in order to quote. Less satisfactory but better than nothing if you do not know your typical utilisation of consumables is to indicate the numbers of WCs, urinals and washbasins on site, and the number of staff.

Additionally, if you intend to include the replenishment of roller towels and the stocking of sanitary towel dispensing machines you will also need to indicate the numbers of each type of dispenser involved.

Typical consumable items that may be included are:

- toilet rolls
- bar soap
- liquid soap
- roller towels
- paper roller towels
- urinal blocks

- sanitary towels
- black bags
- bin liners

Draft something on the following lines:

Consumables

The contractor will be responsible for the cost of and provision of consumable items as listed below. He will also be required to supply, install and replenish roller towels such that the towel dispensers are kept charged at all times. The contractor will be responsible for the laundering of used roller towels.

The contractor will also be required to replenish supplies in the sanitary dispenser machines. It is expected that the contractor will purchase stock from the revenue generated by the machines at no net cost to Neptune Stereophonics. The vend charges must, nevertheless, be agreed with Neptune Stereophonics at the commencement of the contract.

The consumable items required together with average monthly consumption details are as follows:–

	Avg. consumption per month
Toilet rolls white 2 ply, brand X	n_1 *rolls*
Bar soap brand Y	n_2 *bars*
Liquid soap brand Z	n_3 *litres*
Urinal blocks	n_4 *blocks*
Black bags	n_5 *bags*
Bin liners	n_6 *liners*
Roller towels	n_7 *rolls*

There are n sanitary towel dispenser machines on site.

The cost of each item is to be shown on page p of the Tender Return.

Equipment

In the Tender Return document you will ask the contractor to list the equipment he intends to provide. A suitable note needs to be drafted.

Equipment

The contractor is expected to provide the most modern cost effective equipment available. A list of this equipment must be shown on page p of the Tender Return.

Waste Management

The complexity of your notes relating to waste management will depend upon the complexity of your waste management requirements. At their simplest they may read:

Waste Collection and Disposal

All wastebins must be emptied into black polythene bags of a suitable gauge (min: 200 gauge). These bags and white bin liners are to be provided by the contractor and costed with the consumables on page p of the Tender Return. All white liners must be replaced when soiled.

Full black bags containing waste are to be taken to the compactor area adjacent to the Goods Inwards warehouse.

Waste must be removed from wastebins only. Papers adjacent to wastebins or left on the floor must not be treated as waste.

There are n sanitary disposal units on site which should be changed weekly. The contractor is required to show a separate price for the maintenance of these units on page p of the Tender Return.

If however you produce a wide variety of wastes which need to be processed in different ways, your waste management notes will be more complex. The specimen waste management schedule for a manufacturer of household detergents and cleaning chemicals as shown in Figure 5.5 might require notes along the following lines for example:

Fulcrum Industrial Detergents Ltd. are committed to a responsible attitude to waste management and are continually striving to achieve effective reclamation of recyclable materials. The waste management schedule is intended to reflect this policy.

There are two primary objectives:

i.   To provide an efficient waste collection service to the manufacturing areas.

ii. To recycle as much waste as possible.

Typical amounts of waste produced on site per month are as follows:

Waste type	Origin	Average tonnes per month	Suitable disposal arrangements
Sludge	Oils, greases, fats and polyelectrolyte arising from effluent pretreatment	n_1	Disposal via water authority
Non-flammable materials	Reject packaged product or customer returns	n_2	mutration and landfill at approved site
Inflammable product	Reject hydrocarbon filled aerosols or customer returns	n_3	incineration at approved plant
General waste	Canteen waste, soiled cardboard and polythene, scrap timber	n_4	landfill at approved site
Scrap metal	Workshops, drinks cans	n_5	recycle
Cardboard	Packaging	n_6	recycle
Paper	Office stationery	n_7	recycle
Polythene	Wrappers, box liners	n_8	recycle
Laboratory Waste	(a) Sharps, glass	n_9	landfill at approved site
	(b) Solvents	n_{10}	incinerate at approved site

Etcetera

A price per tonne for the disposal of each of these wastes is to be shown on page p of the Tender Return.

Because certain wastes have an intrinsic value it is expected that collection and disposal of these wastes will be free of charge or will have a net value which is to be credited against the total contract price.

We will take this exercise no further except to consider some of the issues raised in the above note, and some which might be raised if it was to be undertaken.

Issues Raised

1. Fulcrum Industrial Detergents have made a brief policy statement in this *General Note* – 'We are a Company under public scrutiny with a reputation to maintain. We want to appear as 'green' as possible.'[1]

2. Each different waste has been designated according to type; classified in terms of its origin or chemical constitution; quantified for costing purposes; and suitable disposal arrangements have been indicated. It might even nominate acceptable sites (where waste is currently being handled for example).

3. The concept of *income* from waste has been alluded to 'not only do we expect you to dispose of our waste but if there is a net return, we expect to get it'.

Issues which might be raised:

1. Equipment. If compactors, mutrators, skips, Eurocarts, racking for storage etc. is required – who will provide it? Certain items e.g. compactors, mutrators, fork lifts and tugs represent a high capital outlay which the contractor may not be able to meet.

2. Separation of waste. Who will sort the waste? The contractor? Personnel on production lines? If the latter, how will it be sorted? Presumably into colour coded Eurocarts for example. If this is the case then the colour code will have to be detailed in these *General Notes*.

3. Keeping adequate records: The need to keep 'cradle to grave' records must to be detailed so that there is no doubt where the first responsibility lies.

If you produce typical office wastes but are inclined towards recycling, remember that you will need to segregate your wastes and will have to advise the contractor of your segregation policy.

Grounds Maintenance

If your grounds maintenance schedule is complex, you may need a set of comprehensive notes to support it. Otherwise simply provide an outline of your requirements. Include:

i. Suitable times for the work to be carried out.

ii. An indication of who is to provide equipment, you or the contractor.

---

[1] This rather complex waste management schedule will not be discussed again and does not appear in Chapter 9. *Preparing the Tender Return Forms.*

iii. An indication of who is to provide consumables – weedkillers, fertilisers, pesticides.

iv. Reference to colour coded site plans if provided.

v. Plant replacement/additions policy.

SUMMARY

The *General Notes* supplement the cleaning schedules with additional information the contractor needs in order to interpret your requirements and price the contract. The information includes:

- notes on areas to be cleaned
- times during which the cleaning may be carried out
- an explanation of the Schedule of Accommodation
- more detail about your periodics and special operations
- an estimation of your consumables usage

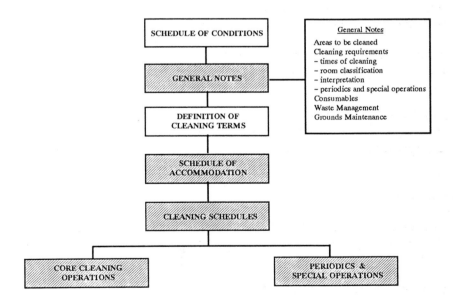

Table 6.1   Factors that need to be considered when drafting the General Notes

1.	Assuming that most cleaning will take place out of normal 'office hours' are any 'day cleaning' operations required?
2.	Will VDUs, facsimile machines, shredders and other 'hi-tech' equipment be included in the specification?
3.	Will telephones be cleaned by the main contractor?
4.	Are there special circumstances relating to the cleaning of computer rooms?
5.	Are there any rooms with restricted access?
6.	Are external areas to be included?
7.	Will deep cleaning of sanitary fittings be required?
8.	Who will be responsible for deep cleaning of the kitchen?
9.	Will carpet cleaning be the subject of a different contract?
10.	What arrangements will be made for wall cleaning?
11.	What arrangements will be made for ceiling cleaning?
12.	Are the light diffusers to be cleaned?
13.	Is window cleaning to be included or will it be the subject of a separate specification? Will the window cleaner be responsible for internal glass partition cleaning?
14.	Is there any confidential waste? What are the waste removal requirements? Is waste to be recycled?
15.	Should storerooms be cleaned?

7

SETTING OUT THE SCHEDULE OF CONDITIONS

The **Schedule of Conditions** lays down the rules with which the contractor is expected to comply throughout the tendering procedure and, if successful, for the term of the contract. As such, it specifies the more formal, contractual aspects of the agreement. It is likely to include details of the following:

- The period over which the contract is to operate.

- The responsibility of the contractor to acquaint himself with all prevailing conditions at the site.

- Disclaimers absolving the client of any errors or omissions in the detail of the specification.

- Contractor's guarantees.

- Staffing requirements at management, supervisory and operative level.

- Security arrangements.

- The need for compliance with prevailing legislation including Health & Safety and the Electricity at Work Regulations.

- The contractor's responsibility to provide materials and equipment.

- Environmental issues.

- The requirement for the contractor to specify his *modus operandi* for the completion of periodic tasks.

- Inspection, liaison and monitoring.

- Quality assurance and quality control.

- Commitment to training.

- Housekeeping arrangements.

- Dress.

- Action to be taken in the event of a fire.

- Action to be taken in the case of non-compliance or breach of contract.

- The client's right to vary the contract.

- Insurance.

- Payment terms.

- Termination of the contract.

- Confidentiality.

As in Chapter 6 we will consider each of these aspects in turn. In many instances, though not all, an illustrative note from the specification for Neptune Stereophonics will appear in italics. Certain conditions fall into distinct categories and thus tend to be linked. In such cases it is preferable to consider these together and to detail them consecutively in the document. Otherwise there is no recommended order in which they should be presented. However, the Introduction clearly comes first.

Introduction

The introduction does two things.

1. It confirms the address to which the specification refers. It is particularly important to include this detail for two reasons.

    i   It removes any risk of ambiguity by clarifying precisely where the work is to be carried out.

        This statement is not as silly as at first sight it may seem. It is not unusual for negotiations for a cleaning contract to take place at the head office of an organisation who may own several buildings. For example at an airport it might apply to one terminal out of several. For a local authority it might apply to the Town Hall, or the City Museum. Yet the negotiations could well take place elsewhere.

    ii If several buildings are to share the same specification it allows each building to be named individually.

        For example it might apply to a local authority's ten fire stations or a bank's sixty branches in a particular region. Each fire station or bank would be listed in this introduction.

It notifies the contractor that you may wish to change the use or internal layout of your site without the need to re-tender or renegotiate a price.

Roger Blackburn prepared the following introduction:-

Introduction

This specification details the cleaning requirements of Neptune Stereophonics plc at 157–173 Prince Consort Road, Chortlon–cum–Hardy, Manchester M21 OAB. The specification refers to the circumstances which currently prevail at the site. Some alterations may occur from time to time during the period for which the contract is extant. In these cases the contractor may be required to vary the type of service supplied. Such variations will be costed in accordance with the variation rates quoted on page p of the Tender Return document.

Definitions

Like any insurance document, hire purchase agreement, lease, mortgage, contract of employment or any one of a multitude of contractual agreements, it is customary to define the parties to the contract. This removes the risk of ambiguity in subsequent sections and obviates the need to use more words than necessary when referring to each party throughout the entire document. Typically you will require two or three definitions:

Contractor: This is the company, firm or person who contracts to provide the services detailed in the cleaning specification. The definition is usually broadened to include the contractor's employees or agent or anyone with whom he sub-contracts to carry out some of the work on your behalf. Make sure you include reference to subcontractors in this way since such operations as window cleaning, grounds maintenance, sanitary towel disposal, provision and laundry of loose laid entrance mats and supply of roller towels might well be subcontracted.

Neptune Stereophonics: Indicate your company name and registered office. You may also wish to say *hereinafter referred to as Neptune*.

The third possibility is:

Representative: i.e. your representative be it an employee of your company, or your designated agent who may be a consultant or part of a contracted management team.

Term of Contract

You have to specify a start and finish date for the contract. First, let us consider the start date.

There are a number of ways in which you might approach this question. It may be for example, that your existing contract has a finite finish date. In this case you must either begin your planning soon enough to terminate the contract on its formal finish date in order to begin a new one the next day; or you must negotiate an extension to give you the time you need to prepare for the new contract.

It may be that you have had the present contract so long that there is no fixed date on which it finishes. In which case you probably have some agreement that allows you to terminate the contract after a period of notice. You must therefore now give this notice.

Whichever applies, the procedure of going out to tender will occupy a significant timescale. Your timetable will need to accommodate the following:

1. *The preparation of your cleaning specification*

This will depend upon the size of your building and the amount of time you are able to devote to the task. By now it should be clear that writing the specification is a time consuming operation.

2. *The issue of invitations to tender*

It may be that you intend to write to a number of contractors to invite them to quote, or you may plan to place an advertisement in a trade magazine, local or national newspaper. If you are in the public sector you may be legally obliged to advertise to issue your invitation to tender in compliance with prevailing EU legislation. These different options will occupy different lengths of time and the decision to advertise in a trade magazine that is a monthly publication will clearly take the longest (unless you have to advertise in the European Journal which will take longer still). In the case of trade magazine advertising there may be a lead time of up to one month before the advertisement appears. Then you will need to allow perhaps another month before the closing date. A third month may well elapse whilst you set up a date for the contractors to make the site visit.

3. *A period for the contractors to submit their tenders.*

If you want the job doing properly you will have to provide enough time for them to complete your tender return document with accuracy. Give them at least three weeks from the date of the site visit.

4. *Scrutiny of the tender return*

Checking each of the tenders and selecting two or three contractors to interview; notifying them of your intention to interview them; and carrying out the interviews will take at least three weeks.

5. *Appointment of the successful contractor*

After the interviews you may require a period for post tender negotiations – fine tuning of some of the prices or arrangements. At the same time the successful contractor will need time to set up the contract, hire the staff, buy the equipment. At least four weeks will elapse.

Figure 7.1 summarises this timetable. 23 weeks have passed by! But at least you now have some idea of the date on which you can schedule the contract to start.

Deciding the finish date is a little easier. Typically contracts operate for 2 or 3 years. Some run for twelve months, or perhaps eighteen, more exceptionally they can be negotiated for a five year period.

One year is too short. If you intend to spend 23 weeks setting up a contract it ought to last longer than 52 weeks. Even if you avoid the lengthy procedure of advertising in a trade magazine, it is quite likely to take four months from the time you go out to tender to the time the contract begins. You cannot afford this expenditure of time in every twelve month period.

Five years is too long. You need to test the market more often. A five year cleaning contract is like a five year parliament. After an initial period of energy and ideas, complacency begins to take a hold, especially in the middle years. By the time the contract comes to an end the faces at the top have all changed and the original objectives have long since been forgotten.

Two years is ideal. So too is three, provided that you build in a formal review and the facility to break after only two.

If you elect for a period longer than one year then you must make it clear that the contractor is required to provide a cost for each year of the contract.

This contract will commence on 1 April 199X and terminate on 31 March 199X+2 subject to earlier termination in accordance with the conditions referred to hereinafter. The contractor will provide a fixed cost for each of the two years of the contract.

You may also wish to allow a probationary period for the contractor to meet the requirements of the specification.

The contractor will be allowed two months from commencement to reach the standards specified.

Pre-Tender Site Visit

For the contractor to make an effective quote, he will need the following:

1. Your detailed cleaning requirements. These are all contained within the cleaning specification.

2. Your tender return document which he is to complete in order to provide the necessary information for you to evaluate his bid. This will be discussed in detail in Chapter 9.

3. An opportunity to familiarise himself with the site. In this respect you will need to lay down some rules and provide some explanation of what is expected.

There are a number of ways in which the visit may be arranged, although only one is preferred.

Some invite contractors to visit the site one by one. This has two significant disadvantages. First, if you have invited five or six companies to quote (which is typical), walking each one round individually is a waste of your time. Secondly, in order to be completely fair in the tendering process then any answer you may give to a question from one contractor should be disseminated amongst all other contractors. If you take them round one at a time, you will need to write to them all with the answers to any questions that have been raised.

Others split their choice of six into two groups of three and escort only two parties around the site. This is far less time consuming than taking them round one at a time, but does not solve the problem of equally sharing all information.

Thus, the best format for arranging a site visit is to invite all contractors to visit at the same time. Nevertheless you may then have a logistical problem of trying to conduct perhaps eighteen people around your building.

The first part of your note may read:

It is a condition that contractors wishing to tender must visit the premises to acquaint themselves with all aspects of the site and any restrictions which may affect their ability to perform the work detailed in this Specification. Any claims which may subsequently arise because the contractor has not satisfied this condition will not be entertained.

It is appropriate next to refer to the floor plans and schedule of accommodation in order that in the same context you can disclaim responsibility for errors as discussed in Chapter 3 and typified by figure 3.9.

Internal areas which are to be cleaned are identified on the site plan which forms part of this Specification. Additionally, each room or area which requires a cleaning service is detailed on the Schedule of Accommodation. This Schedule additionally lists floor areas which have been recorded in square metres.

Whilst every effort has been made to provide accurate information with respect to floor areas, Neptune Stereophonics cannot be held responsible for any errors or omissions. It is therefore the contractor's responsibility to confirm that these areas are correct in order to price the work.

Any discrepancies should be brought to the attention of Neptune Stereophonics.

The next series of conditions is concerned with staff, materials and equipment i.e. the logistics of the contract and working practices. The first of these is:

Staffing Requirements

Ideally as is described in Chapter 10 you will be calculating the manpower requirements necessary to meet your specification. The reason why you need to do this is to check the contractor's estimates of manning levels in order to satisfy yourself that he intends to provide sufficient effort to meet your needs. You will not inform him of your own estimates however, for as we shall see later, in the tender return documents you will ask the contractor to show how many hours he calculates are necessary. Suppose also (for illustration), your specification requires 300 hours per week for core cleaning operations. Then you will require 300 hours *every* week. If one person working a 30 hour week is off sick, and another is on holiday, there will be a 20% shortfall in effort. *20% of your scheduled core cleaning operations will not be done.* One of the conditions you lay down must therefore stipulate that sickness and holiday cover are to be provided. General absenteeism on a day to day basis must also be accommodated. You may wish to draft terms along the following lines:

Manning levels as indicated in the Tender Return document, must be maintained at all times. This means that the contractor must ensure that arrangements are made to provide additional labour to cover sickness, holidays or general absenteeism on a daily basis. The contractor may also be required to provide additional staff if requested to do so by a representative of Neptune Stereophonics.

Under this same subheading, you may also wish to comment upon your expectations with respect to supervision. There are a number of options.

1. A site manager may be appointed by the cleaning contractor. In such cases he or she will work exclusively on your site.[1] You cannot expect a site manager to be provided if yours is a relatively small site. The site manager will be involved in the hiring and firing of the workforce; the ordering of consumables and cleaning materials; and will be your principal point of contact with the contractor. You may wish to stipulate minimum requirements for qualifications and experience.

2. The workforce may only be supervised. In such cases the so called *non-working supervisor* may well become an extra hand in the event of absenteeism or sickness, which is not desirable. The actual management will be off-site with the contractor's representative looking after several contracts simultaneously and spending their time equally between all of the sites they manage.

3. The workforce will report to a *working supervisor* i.e. a senior cleaner who is responsible for both local 'management' and for cleaning part of the building themselves. Such arrangements are typical at sites with only six or so cleaners.

Whatever is appropriate on your site you will need to draft a suitable comment. In the case of a site manager you may wish to say:

The contractor shall provide a site manager, acceptable to Neptune Stereophonics, whose sole responsibility is management of the cleaning at the Prince Consort Road site. The manager must be on site during the main cleaning shift from 2300–0700 Monday to Friday but must also be available for regular, scheduled meetings with the nominated representative of Neptune Stereophonics. Day cleaning, periodics,

---

[1] Consider giving them a desk in an open plan area amongst your own personnel. In this way they begin to identify with your own corporate identity rather than their employers. It becomes *their* building that they are keeping clean. Not simply their client's.

special operations and core cleaning to be carried out outside the hours of 2300–0700 or at weekends shall be suitably supervised. It is the contractor's responsibility to identify the level of additional supervision felt necessary to ensure that cleaning is being carried out in accordance with the specification and that staff are working correctly. Details of management and supervisory staff must be shown in the Tender Return document.

In order to support the foregoing paragraphs some additional detail with respect to site management is advisable. You may wish to include it under the same subhead **Staffing Requirements**, or introduce a new section as follows:

Staff Contacts

The contractor will notify Neptune Stereophonics of the name of their staff contact. Any change in personnel or in this arrangement will be notified to Neptune Stereophonics at once.

Neptune Stereophonics will nominate a representative who will be responsible for all liaison with the contractor. The contractor will be advised in writing if the nominated representative is to be replaced or otherwise changed.

Except in the case of independent audits (q.v.) the nominated representative of Neptune Stereophonics alone shall judge whether the contractor's performance meets the requirements of the Specification.

You may also wish to include in this section a requirement for the contractor to make available on a monthly basis, a senior management representative who will formally visit you to discuss whatever problems you may have.

Contractor's Employees

You will need to consider this aspect of your *Schedule of Conditions* very carefully. Many lessons have been learned by the naive or unsuspecting and some of the most respectable cleaning contractors have had problems in the past following raids by Immigration officials or by Social Services inspectors. Today contractors are ever conscious of the risks associated with employing casual labour and carry out whatever checks can reasonably be expected. Nevertheless some illegal immigrants, tax dodgers or social security cheats still manage to find employment and occasionally get caught. When that happens the client's name inevitably finds its way into the newspapers especially if the name is well known. At the very least you should set down in your *Schedule of Conditions* the terms under which you will accept your contractor's employees. You may want to include the following:

i. That each employee should complete an application form which shows their full name, address, previous employment and National Insurance number.

ii. That the contractor shall check original DSS documents to confirm that the National Insurance number is valid and genuine.

iii. That the contractor shall check passports of non EU nationals to ensure that they have a valid entry visa and work permit.

Whilst each of these relate to the legal right to work, this is the place to add supplementary conditions concerning the workforce. For example:

iv. That the contractor shall seek references for every employee.

v. That you reserve the right to reject any employee of the contractor if you consider them to be undesirable.

vi. That the employee has no convictions other than spent convictions as defined by the Rehabilitation of Offenders Act 1974 (except that employers are not now permitted to enquire about an employee's past criminal record).

vii. That all employees must be able to understand and converse in English. There are many examples where this is not the case. Failure to be able to understand English threatens Health and Safety requirements and prejudices training and supervision.

viii. That all employees should sign an attendance register. Ideally this should be under the supervision of your security staff or some third party. Not only does it enable you to confirm that labour levels meet the requirements of the *Specification* and the promise of the contractor in the *Tender Return* but it also provides a record of the number of people on site in the event of a fire. (Watch out for cheating!).

ix. That the contractor shall provide you with a list of names of all personnel working on your site and should keep that list up-to-date by notifying any changes as they occur.

x. That the employee must agree to be searched if required.

Within this same sub-section, you may wish to add:

xi. That failure by the contractor to comply with any of these requirements may result in penalty or termination of the contract.

xii. That you will hold the contractor responsible for the actions of his employees or any subcontractor.

xiii. That the contractor must ensure that any employee, agent or subcontractor found to be under the influence of alcohol or drugs shall be removed from the premises.

xiv. That the contractor is solely responsible for supervision, control, payment of wages including the withholding of income tax and social security payments, redundancy payments, compensation or unfair dismissal payments, disability benefits, sick pay, holiday payments etc.

xv. That at your discretion, and with reasonable grounds you may require the contractor to replace employees within a stipulated period of time and that under exceptional circumstances you may require an employee to be removed at once. Under these circumstances you will expect the contractor to find a replacement as soon as possible.

xvi. That the contractor's contract with his employees is such that he is able to comply with your conditions of contract.

xvii. That you reserve the right to examine the contractor's personnel records.

Equipment and Materials

Unless you have decided otherwise it will normally be the contractor's responsibility to ensure that he has provided all of the necessary equipment and materials to meet your specification. Indeed he will be expected to list the equipment he intends to buy and the materials he plans to use, when completing the tender return. Make this clear in your *Schedule of Conditions*. Also make it clear that he should not provide materials or equipment, other than those specified, without your knowledge and consent.

You will also wish to absolve yourself of any loss of or damage to his materials and equipment, making it his duty to take measures appropriate to its security. Require him to report any incident of theft, tampering or damage to your own security staff. However, you have a responsibility to provide some lockable storage facility to make this easier for him. (q.v.) Stipulate that all equipment and materials shall be returned to store at the end of each shift.

Set out a condition that the contractor must clearly mark his equipment for identification of ownership; that at all times it must be clean and presentable and in good working order; that it is safe; and that it is checked in conformance with the Electricity At Work Regulations 1989.

Stipulate that only trained personnel may use electrical equipment and that this must be disconnected from the mains at the end of each working shift. Point out that no electrical equipment belonging to you is to be unplugged by any member of the contractor's staff for the purpose of using cleaning equipment.

Indicate that the contractor will be responsible for any damage caused to the fabric of the building, or furniture, fixtures or fittings as a result of misuse of equipment or materials; or by negligence whilst carrying out his work. It is not uncommon to see bleach damaged carpets or stains from window cleaner's buckets for example. Make the contractor liable for any consequential loss.

You should also consider whether the contractor needs your authority before he takes delivery of materials and equipment. Normally you would expect his staff to be responsible for receiving his incoming materials, for checking them off against a delivery note, and for manhandling them to his central store.

In the environmentally conscious nineties indicate your commitment to the use of environmentally friendly products that have no depleting effect on the ozone layer and are biodegradable.

Contractor's Accommodation

You will need to provide the contractor with both storage facilities and office accommodation.

If you operate from a large site, or on several floors you may need to provide several lockable rooms or cupboards for storage. Coat lockers on the fifth floor, just capable of holding a mop and bucket are not enough. Nor is it acceptable to expect your contractor to store his equipment behind the door of some less frequently used toilet. He will need at least one main storeroom where he can keep large items of cleaning equipment and forward supplies of materials. This room will also have a low sink for filling buckets and if no additional facility is available, will contain clothes lockers for cleaning staff to use. If, as at Neptune Stereophonics, cleaning operatives work shifts long enough to qualify for a break then a 'Mess Room' of some type should also be provided unless of course the main store is large enough for it to be located there. The contractor will also require a lockable room to maintain supplies of the consumables he stocks on your behalf.

Elsewhere in the building or buildings, cleaners cupboards need only be large enough to store the equipment they need though ideally these too should be fitted with low sinks.

You will also need to provide office accommodation for the site manager or non-working supervisor – whichever you have. As has already been noted you may want their 'office' to be a desk in an open plan area occupied by your own staff. However, if this is the case they will also need access to some other area for interviewing, disciplining or otherwise having confidential meetings with their staff or superiors. They will also require a telephone which should be provided free of charge. You would normally pay the rental but expect the contractor to pay the call charges. This may be difficult to police, just as it is difficult to ensure that cleaners do not make long distance or international calls when you are away from the site. However, as the case of Heasmans v Clarity Cleaning Co. Ltd. (1987) showed, it is your responsibility to ensure that the cleaning staff do not abuse your telephone system, not the contractor's.

With these considerations in mind you will draft something along the following lines:

Neptune Stereophonics will provide office accommodation for the contractor's use. A telephone will also be provided. Calls will be metered and charged at cost. The cost of telephone line rental will be borne by Neptune Stereophonics.

Lockable storage facilities will also be made available around the site. All of these stores must be kept in a clean, tidy condition at all times. At the end of each shift all contractor's equipment will be cleaned and returned to its designated storage point.

Within this subheading you may also wish to introduce some aspects of fire and material safety. For example:

No flammable materials shall be used on site.

Or:

No flammable materials shall be used without obtaining approval from Neptune Stereophonics. In the event that this approval is granted, such materials shall be stored in fire resistant cabinets at all times.

With regard to material safety you may add:

The contractor will be required to provide, in advance, material safety data sheets for every product he intends to use on site. A copy of these data sheets will be held both by the Safety Officer of Neptune Stereophonics and by the contractor's Site Manager. (or Supervisor)

Periodic Scheduling

This section draws to a conclusion the series of conditions relating to logistics and working practices.

Monitoring of core cleaning operations is relatively easy because the tasks are carried out frequently enough for it to be obvious if they are missed. However, periodic operations are more difficult to monitor subjectively. Furthermore, some periodics are disruptive and you may need notice before they are undertaken. Additionally as discussed in Chapter 5 you will only be paying for periodics upon completion. Each of these aspects need to be covered in what you draft:

Upon appointment, the contractor will be required to schedule for each year of the contract, all tasks which are to be performed less than monthly. This schedule, will be drafted in consultation with Neptune Stereophonics, and may only be amended with their approval.

Periodic tasks must be identified separately on submission of each monthly invoice and will only be paid upon presentation of a satisfaction note, signed by the appointed representative of Neptune Stereophonics.

The next series of conditions is generally concerned with performance and quality control.

Training

Training procedures in the cleaning industry are generally poor, especially at operative and supervisory levels. There are a number of reasons.

i. Some operatives do not speak English (hence the condition earlier in the section headed *Contractor's Employees*). This means that unless their supervisor speaks their language a real barrier may exist.

ii. Labour tends to be itinerant. Turnover as high as 150% per annum is reported by contractors in some areas.

iii. Peer group training is common. Unfortunately this presumes that other members of the group have been adequately trained. Usually they have not. There is thus a tendency for bad practices to be passed on from employee to employee.

iv. There is no incentive to become skilled.

During interview, just before you appoint your contractor, most will show you elaborate, well produced training aids including manuals, slide presentations, videos, cartoon strips and the like, all of which they claim are used for training purposes. Usually however, this training stops at supervisory level. The key issue must therefore revolve around how successful is the supervisor in disseminating this information to the operatives? If you expect too high a result you may well be disappointed.

You must therefore draft conditions which set out your requirements for training. At the very least you must demand that operatives are given some form of induction to ensure that they are conversant with specific aspects of your site including treatment of special finishes, signing in procedures, fire regulations, minimum standards etc. Supervisors should have limited management skills at least to 'team leader' competence. They should also be completely familiar with your specification.

Indicate that you expect the contractor to have a responsible attitude towards training and to keep you fully aware of his programme and the level of competence achieved by each of his staff.

Quality Assurance

The contractor must expect that you will require him to have some form of Total Quality Management system in place. This may be BS5750 : Part 2. It may be an alternative.

There is an increasing trend to specify that if the contractor has not yet been registered by the British Standards Institution he should be striving to achieve registration. If you are prepared to accept this then why not set a date by which you expect he will have achieved certification.

Do not forget however that there are Quality Management systems other than that specified in BS5750 : Part 2 (equivalent to ISO 9002). If a contractor can convince you that his procedure is as good (or better) you may be perfectly satisfied to accept his own in-house alternative.

Whatever system is in place however, you will need to ensure that effective quality control is being implemented. Make it clear that you expect the contractor to monitor performance on an ongoing basis. Reserve the right to inspect his quality control records. For example:

It is a condition that the contractor shall comply with the requirements set out in BS5750 : Part 2 and that this compliance is formally registered by a suitable independent body In some circumstances a contractor may be appointed even though he has not yet fully met all the requirements of the standard. In such circumstances he will be expected to do so within twelve months of the commencement of the contract. Failure to do so may result in early termination.

Notwithstanding his current position with respect ot BS5750 : Part 2 it will be the contractor's responsibility to monitor the quality of the work performed and to ensure that it meets the standards provided by the Specification. Neptune Stereophonics reserve the right to inspect quality control documentation at any time without notice.

At the time of writing it has been claimed that public sector organisations cannot demand compliance with BS5750 although this position has not yet been legally tested. In any case changes in EU legislation have now made significant differences to the way in which public sector contracts are awarded by tender.

Monitoring

You will want to monitor the performance of the contractor throughout the life of the contract. You may wish to do this yourself. Alternatively you may wish to retain an independent auditor. There are a number of checks to be carried out.

i. Confirm that all items that are supposed to be cleaned daily are cleaned daily. The build up of dust for example is an obvious indication that a cleaning task is being missed.

ii. Confirm that the number of operatives on site matches the number of operatives indicated in the *Tender Return*.

If you do not have security staff on duty who keep a record of the persons who are on site you should not rely on the cleaning contractor's own signing-in procedures. People have been known to sign on behalf of their friends. Why not activate the fire alarm at 0630 – or whatever time the cleaning staff are on site – and count them as they come out of the building. You may be surprised. There is one documented case where 18 cleaners assembled in the car park when the Tender Return indicated that 33 were necessary to meet the requirements of the specification! (There were in fact about 28 signatures in the attendance book). If your specification demands 33 cleaners but only 18 are in attendance you should only be paying 18/33 rds of the contract price. Do not accept the concept that any shortfalls will be made good next day. If the ashtrays need to be emptied daily but are overlooked, it is no use emptying them twice tomorrow!

iii. Confirm that the number of supervisors on site matches the number of supervisors indicated in the *Tender Return*.

iv. Confirm that the wage rates indicated in the *Tender Return* are indeed being paid to the operatives and supervisors.

v. Check that vacuum cleaner bags are empty and that all equipment is in good working order.

vi. Check that records relating to the Electricity at Work Regulations 1989 are being maintained.

vii. Check that Material Safety Data Sheets are available for all chemicals in use on site and confirm that the chemicals are those specified in the *Tender Return*.

viii. Confirm that all cleaning staff are wearing the specified uniforms.

ix. Check that operations are being carried out as defined in the *Definition of Cleaning Terms* section of the specification. In particular note that no dry dusting is being carried out when damp dusting is intended and that feather dusters are not being used.

If you intend to subject the contractor to this scrutiny you must make this clear in the *Schedule of Conditions*. The following example may form a suitable basis:

Neptune Stereophonics will regularly monitor the performance of the contractor in relation to the requirements of the Specification and may from time to time use an independent auditor for this purpose.

Such audits will seek to confirm that specified tasks are carried out at the scheduled frequency and in the proper manner as defined in this Specification. They will also compare the actual allocation of labour and the provision of equipment and materials against those indicated in the Tender Return.

You may also wish to make reference to penalties that may be invoked if monitoring reveals a shortfall in performance. The ultimate penalty is of course early termination of the contract (q.v.) although this is more likely to result from habitual and serious neglect rather than occasional or superficial noncompliance. However the invocation of penalties is to be avoided if at all possible. It is far better to ensure that adequate quality control is being implemented by the contractor thereby reducing the likelihood of noncompliance.

It is reasonable to allow the contractor's representative to accompany you or your independent auditor whilst any checks are being made and you may wish to include a note to this effect. For example:

The contractor will be given an opportunity to accompany the representative of Neptune Stereophonics at such times as inspections may be carried out. It is in the contractor's interest to do so.

You may next wish to bring together a number of issues relating to safety, discipline and procedural matters intended to ensure compliance with legislation or with your own internal rules. In no particular order these include:

Security

You may well have formal security measures already in place and procedures will therefore be well established. In such cases the *Schedule of Conditions* will simply reinforce your expectations. Typically you may draft:

The contractor will be required to conform to all current security procedures at Neptune Stereophonics.

Before the contract commences, the contractor must supply names and addresses of all personnel he intends to employ. In the event that cleaning staff are replaced during the term of the contract, the name and address of any intended replacement shall just be notified to Neptune Stereophonics.

All contractor's personnel will be issued with identity badges bearing their photograph which they are required to wear at all times. It is the contractor's responsibility to ensure that Security are notified of any changes in personnel and that badges are withdrawn from persons no longer authorised to wear them.

All personal baggage is subject to search on entering and leaving the premises and must be locked away at the beginning of each shift. Neptune Stereophonics also reserve the right to examine all equipment and tools and to search contractor's staff (with their consent). Such tools and equipment must be clearly marked for identification purposes and stored away when not in use.

If the contractor is unable to enter any area that is scheduled to be cleaned, access should be gained from a member of the security staff.

Neptune Stereophonics reserve the right to refuse access to anyone not directly involved in the contract or to anyone who is employed by the contractor, without giving any reason.

Legal obligations

The contractor has a number of legal obligations to fulfil. These include conformance with the following:

    i.   Sex Discrimination Act 1975

    ii.   Race Relations Act 1976

    iii.   Equal Pay Act 1970

    iv.   Health and Safety at Work Act 1974

v. Electricity at Work Regulations 1989

vi Manual Handling Operations 1992

Other legislation concerned with employment of persons under school leaving age; with the Control of Substances Hazardous to Health (COSHH regulations); with discrimination on the grounds of religion, sexual orientation or disability; and with various EU directives, also prevails.

It is the contractor's responsibility to comply whether or not you spell out the need to do so in your *Schedule of Conditions*. You may prefer to do so.

Nevertheless of the above, Health and Safety tends to be the one most commonly detailed as follows:

Health and Safety

As well as reinforcing your position that it is the contractor who is primarily responsible for the Health and Safety of his staff you may wish to take this opportunity to list certain aspects that may, through the contractor's negligence, affect your own staff. Such items include:

i. That all gangways, fire escapes, stairways, ramps and passages are to be kept free of obstruction.

ii. That floors are to be kept free from contamination with oil, water, spillages in general; trailing wires, equipment and tools.

iii. That flammable waste must not be allowed to accumulate except in designated waste disposal areas where proper storage and handling facilities will be provided.

iv. That materials and equipment are to be stored safely and securely.

v. That work in progress or hazard warning signs are displayed wherever wet floor cleaning is in progress.

vi. That fire doors are not to be wedged open and that there will be no tampering with or obstruction of fire fighting equipment.

vii. That all COSHH regulations are met.

You may also wish to demand that the contractor has a nominated member of staff who is trained in first aid.

You may also wish to add:

In the event of any accident, injury, fire or damage to property, the contractor shall at once notify the competent authority within Neptune Stereophonics and shall subsequently confirm this report in writing.

Fire Procedures

You will of course issue all contractor's personnel with details of your fire regulations. Nevertheless you may once more wish to reinforce your position by including some details in your *Schedule of Conditions*. For example:

In the event of the contractor discovering a fire he must immediately activate the fire alarm. The contractor should be familiar with all such alarms which are located extensively throughout the building. (You may wish to say where they are if they are in a particular location on each floor). *The contractor must take no unnecessary risks and must comply with the designated evacuation procedures.*

Upon hearing the fire alarm the contractor should make safe all equipment in use at the time and should then leave the building in an orderly manner by the nearest exit. Once outside he should make his way to the designated assembly point.

A nominated representative of the contractor (Site Manager; Supervisor) *will be responsible for confirming that all persons employed by the contractor, or under his control, have safely evacuated the premises by calling a roll. Day cleaning staff will be under the direction of a Neptune Stereophonics representative.*

Once the alarm has been raised no person may re-enter the premises without the permission of a duly authorised person.

Fire hoses and extinguishers should be used only by trained personnel.

Neptune Stereophonics is a No Smoking site.

If yours is a No Smoking site you may wish to highlight this fact elsewhere, although the section dealing with fire precautions seems as good as any other.

Protective Clothing

Associated with Health and Safety it may be appropriate at this point to continue with your requirements by stating that the contractor shall wear protective clothing. This may simply mean an overall which you may demand for no other reason than to differentiate between your own staff and cleaning staff. It may mean special suits for handling certain types of waste; hard hats; safety glasses; steel toe-capped shoes, etc. Draft something along the following lines:

The contractor shall provide a uniform for his employees bearing the company's logo and to a standard acceptable to Neptune Stereophonics. This should be at no charge to the contractor's own staff. Operatives shall wear footwear suitable for the type of work being undertaken. Open sandals or high heeled shoes are not acceptable. Safety shoes and hard hats shall be worn at all times in the warehouse areas and safety spectacles shall be worn within manufacturing areas. These will be provided by the contractor to Neptune Stereophonic's specification.

Liability and Duty to Insure

You will require the contractor to be adequately insured.

The contractor shall indemnify Neptune Stereophonics against any claim, cost or proceedings against Neptune Stereophonics for any personal injury or death or loss or damage to property caused by the contractor's negligence, breach of contract, breach of statutory duty or otherwise.

The contractor shall indemnify Neptune Stereophonics against all direct (or indirect) losses incurred by Neptune Stereophonics as a result of accidental damage by the contractor or as a result of the contractor's negligence, breach of contract, breach of statutory duty or otherwise.

The contractor shall at all times, maintain proper and adequate insurance cover in respect of all insurable risks including public liability insurance cover with a minimum indemnity of 2 million pounds (or whatever) *in respect of any one claim or incident.*

Evidence of such insurance shall be furnished to Neptune Stereophonics by the contractor on request and such insurance cover shall on no event be cancelled or allowed to lapse by the contractor during the term of this agreement.

Vicarious Liability

Along the same lines:

The contractor shall accept vicarious liability for the misbehaviour of his employees or subcontractors.

There next follows a series of fiscal matters.

Variation to Contract

It may be, during the period of the contract, that you wish to change the use of certain rooms. As we shall see later you will require the contractor, at the time he makes his tender submission, to indicate a cost per 100 m^2 for cleaning each category of room in your classification over a 28 day period. For example, the cost of cleaning an office may be x per 100 m^2; an executive area 2x per 100 m^2; a toilet 3x per 100 m^2 and so on. If you subsequently change 100 m^2 of executive rooms into normal offices you would expect the cost of cleaning that area to be reduced by a factor of 2. Conversely if you converted 100 m^2 of offices into toilets the price would increase by a factor of 3. Alternatively, you may build an extension comprising a mixture of offices, toilets, corridors, production areas etc. The tender return document will provide the formulae to allow you to calculate the changes in costs. All that you need do at this stage is to draft a condition giving you the right to do so. For example:

Neptune Stereophonics reserve the right to vary the contract should the need arise. This may result in a change in the contract price. Such changes will be based upon information supplied by the contractor on page p of the Tender Return document.

Payments

As has been discussed, payments are made up of two components. One is for core cleaning operations which are paid in twelve (or thirteen) equal monthly (or four weekly) payments. The other is for periodics which are paid upon presentation of a satisfaction note.

Decide whether you will pay twelve or thirteen times per annum; Decide upon your payment terms – one, two or three months in arrears; and draft a condition accordingly. Remember to spell out your terms for the payment of periodic tasks. For example.

The contractor will be required to submit an invoice each calendar month. This will detail:

    a.   periodic and special operations completed during the month

    b.   consumables purchased during the month

    c.   costs or income from waste management operations arising during the month

    d.   one twelfth of all other costs as set out in the tender submission

Neptune Stereophonics undertake to make payments thirty days in arrears.

Payment for periodic tasks and special operations will only be sanctioned for payment upon presentation of agreed documentation confirming Neptune Stereophonic's satisfaction with the work undertaken.

Recovery of Sums Due

It is beneficial to have some clause that enables you to deduct from the amount you pay the contractor each payment period, any sum of money which he may owe you for whatever reason e.g. telephone call charges; adjustments relating to consumables, penalties etc. A suitable clause may be:

It is a condition of contract that any sum of money recoverable from or payable by the contractor may be deducted from any sum due to the contractor at that time, or which may become due to the contractor at some later time.

Value Added Tax

Draft a clause which clarifies that you expect the quoted price to be net of VAT which you agree will be added to each monthly invoice at the prevailing rate.

Insolvency

Some prefer to include termination clauses in the event of insolvency or bankruptcy of the contractor as an individual, or as a partnership; or of appointment of a receiver, administrator, winding up order, or possession order in the case of a limited company.

Your condition should demand notification in writing should any of the above events occur and you should retain the right of any action or remedy which may have accrued on your behalf.

Confidentiality

State:

All documents and information which the contractor may receive or have access to in connection with the performance of the contract, are confidential and may not be disclosed to any person without permission of Neptune Stereophonics unless a duty to disclose is imposed by statute or Court order.

Assignment of Contract or Right to Subcontract

You will expect the contractor to appoint subcontractors only with your knowledge and approval. Require that he obtains your consent in writing before subletting, assigning or transferring the contract or any part of it to any third party.

Noncompliance

Indicate the consequences of failure to comply with any of the foregoing conditions as follows:

Failure to comply with any of the conditions set our herein may result in the contractor not being allowed upon the premises. In such circumstances Neptune Stereophonics will not be responsible for any consequential loss.

Finally set out the conditions under which you reserve the right to terminate the contract.

Early Termination

First, give yourself the right to terminate the contract without the need to give any reason. In order to do this you must specify a period of notice. Twenty eight days or three calendar months are usual. The contractor may also ask to have this same right granted, i.e. the right for him to terminate the contract. However, if you agree to anything less than three calendar months notice on his part, you may find it difficult to set up a new contract with a different contractor in the time available. Insist upon three months.

For under-performance write in a clause whereby you can demand in writing that the under performance should be remedied by a due date. Failure to remedy the under-performance should risk the penalty of termination on seven days notice:

Neptune Stereophonics may terminate the contract at any time and without stated reason upon giving 28 days notice in writing.

If the contractor fails to meet the requirements of the specification he will be advised of the failure in writing and be allowed a fixed time in which to remedy it. Failure to do so in the allotted time may result in termination of the contract upon 7 days notice in writing.

Neptune Stereophonics reserve the right to terminate the contract at once in the event of any serious breach of contract or misdemeanour on the part of the contractor.

It may be, during the term of the contract, that you no longer need to occupy all or part of the building. Write in a clause which allows you to vary all or part of the contract for these reasons after giving a stated period of notice.

Your *Schedule of Conditions* is now complete. Figure 7.2 summarises the headings used in this chapter which you may wish to use in your own document.

SUMMARY

The *Schedule of Conditions* formally lays down the rules under which you intend to operate the contract.

The subjects which are detailed broadly fall into the following categories:

- general operation of the contract including term of contract, tendering procedures, staffing requirements

- quality assurance including training, quality control, monitoring

- safety procedures incorporating Health and Safety legislation, fire precautions, COSHH legislation

- right to terminate the contract for unsatisfactory performance or general misdemeanour

- payment terms

Except for one final component the formal specification document is now complete.

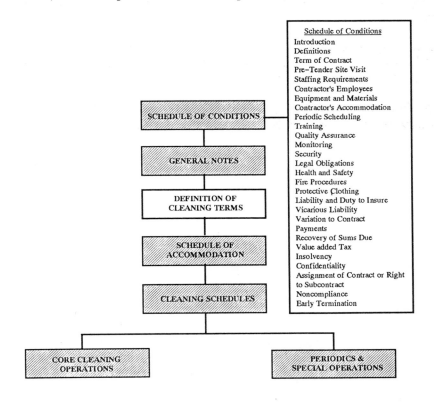

Figure 7.1: *Typical timetable from writing the specification until commencing the contract*

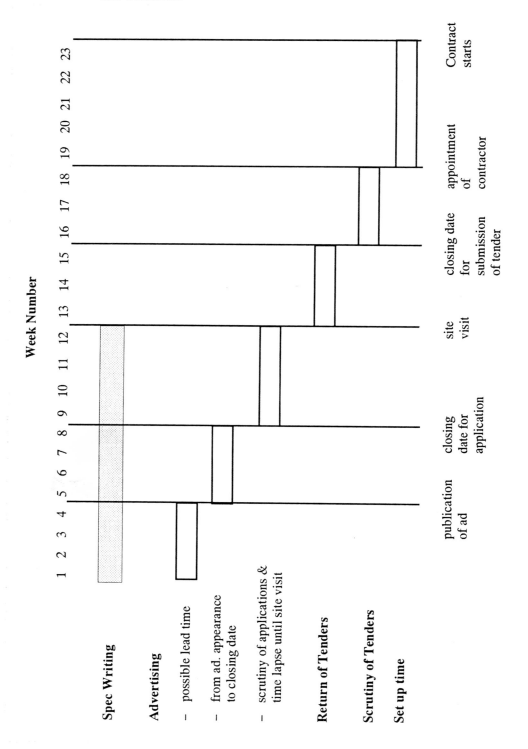

Figure 7.2 Summary of headings to appear in Schedule of Conditions

1. Definitions
2. Term of Contract
3. Pre-Tender Site Visit
4. Staffing Requirements
5. Contractor's Employees
6. Equipment and Materials
7. Contractor's Accommodation
8. Periodic Scheduling
9. Training
10. Quality Assurance
11. Monitoring
12. Security
13. Legal Obligations
14. Health and Safety
15. Fire Procedures
16. Protective Clothing
17. Liability and Duty to Insure
18. Vicarious Liability
19. Variation to Contract
20. Payments
21. Recovery of Sums Due
22. Value Added Tax
23. Insolvency
24. Confidentiality
25. Assignment of Contract or Right to Subcontract
26. Noncompliance
27. Early Termination

8

DEFINING THE CLEANING TERMS

There is a body of opinion that suggests that a series of definitions of cleaning terms as a component of the specification, serves no useful purpose. Those who are of this view argue that once a task has been specified it needs no further qualification and may be left to the preference or discretion of the cleaning operative as to how that task should best be carried out. If you are of this opinion read no further but go directly to Chapter 9.

Roger Blackburn decided to include a section defining the terms in his specification because he knew full well that any subsequent arguments about performance or quality would need some yardstick against which they might be assessed.

Several scientific organisations around the world have been engrossed in trying to determine, to use the well-worn cliche, 'how clean is clean'. It can be done. There are techniques to quantify soil levels on 'smooths' or in carpets and bacterial levels can be assessed by well established techniques. Some have developed *grey scales* or similar subjective procedures, or have measured *gloss* or *reflectivity*. Unfortunately none of these are of real practical benefit. They are of limited use, either because they require the purchase of expensive laboratory apparatus; or they demand special skills of measurement and interpretation; or they are simply too time consuming, yielding a result that is two or three days too late to be of any value.

Thus it may be argued that the inclusion of a glossary of cleaning terms within the body of the specification reduces the chance of argument about acceptable standards at the time any formal monitoring of performance is being carried out.

In the following pages a series of definitions is proposed. You may wish to adapt them to suit your own preference or to import them *en bloc* into your specification. Hopefully, the list will be more comprehensive than you need, in which case those that do not apply can be omitted. If others do not appear in the list draft a suitable definition along similar lines. Remember you should provide a separate definition for every task that appears in your cleaning schedules. In most cases you will note that each definition describes the operation and its objective.

Absorbent powder clean carpets

Apply absorbent powder cleaner in accordance with the manufacturer's recommendations and brush into the pile of the carpet using suitable equipment. Allow to dry. Vacuum to remove residual powder and absorbed soil.

Apply antistatic treatment to carpet

Thoroughly vacuum the carpet and pre-spot as necessary. Apply a proprietary antistatic agent from a suitable sprayer in accordance with the manufacturer's instructions. Wipe off all overspray from furniture, fixtures and fittings.

Bonnet Buff
Bonnet Clean

Thoroughly vacuum the carpet using a twin motor upright machine and pre-spot as necessary. Remove superficial soil using a rotary carpet cleaning machine operating at a speed not exceeding 175 rpm and fitted with a 'bonnet' pad impregnated with a suitable chemical. Realign pile using a pile rake if required.

Buff
Burnish

Using a high speed (300-500 rpm) or ultra high speed (1000 rpm or more) rotary machine fitted with a suitable brush or pad, buff/burnish the floor to achieve a shine free from scuff, abrasive and heel marks. If the floor has been polished the machine should be fitted with a suction unit or else the floor should be swept or dust control mopped after the buffing/burnishing operation is complete.

Change curtains
Change net curtains

Take down curtains and remove for laundering. Launder in accordance with the manufacturers instructions. Clean curtain track to remove dust and dirt particles. Rehang replacement curtains or laundered curtains as available.

Clean

Wipe using a suitable cloth impregnated with a proprietary cleaning agent to remove dust, dirt and superficial marks. If necessary, polish with a dry lint free cloth or paper towel to remove cleaning marks.

Clean blinds

If the blinds are Venetian blinds:

Wipe using a suitable cloth impregnated with a proprietary cleaning agent to remove dust, dirt and superficial marks. Wipe dry using a lint free cloth to remove cleaning marks.

If the blinds are vertical blinds:

Vacuum clean with an appropriate tool. Wipe with a suitable cloth impregnated with a proprietary cleaning agent to remove any marks that remain after vacuuming.

Clean and deodorise drains

Remove grids and gully covers and clear channels of debris. Scrub grid and channel using a stiff brush or powered brush with a proprietary drain cleaning product. Flush away all residues and deodorise as necessary using a proprietary deodoriser.

Clean and polish dry stainless steel
Clean and polish dry metalwork
Clean and polish dry brasswork

Wipe using a suitable cloth impregnated with a proprietary cleaning agent to remove dust, dirt and superficial marks. Polish dry with a lint free cloth to remove cleaning marks. Do not use abrasive cleaning creams.

Clean ceilings

Depending upon the type of ceiling, either:

Wipe using a suitable cloth impregnated with a proprietary cleaning agent to remove dust and dirt, cobwebs and fluff.

or;

Vacuum clean with an appropriate tool to remove dust, dirt, cobwebs and fluff.

Clean ceiling vents

Vacuum clean with an appropriate tool to remove loosely adhering dust and dirt. Wipe using a suitable cloth impregnated with a proprietary cleaning agent to remove residual soiling.

Clean door glass
Clean glass in furniture and fixtures

Wipe using a suitable cloth impregnated with a proprietary glass cleaning agent to remove dust, dirt, fingermarks and other superficial marks from the entire surface and frame. Polish dry with a lint free cloth to remove cleaning marks.

Clean door kickplates, pushplates and handles

Wipe using a suitable cloth impregnated with an appropriate cleaning agent to remove dust, dirt and superficial marks from door kickplates, pushplates and handles. Polish dry with a lint free cloth to remove cleaning marks.

Clean glass

Use suitable window cleaning equipment and an appropriate detergent to remove dust, dirt, fingermarks and other superficial marks, changing the detergent solution with sufficient frequency to avoid contaminating the surface being cleaned with dirty water. If necessary polish dry with a lint free cloth to remove cleaning marks. Ensure that all adjacent surfaces such as sills, mullions and frames are left free from smears.

Clean interior surfaces of cupboards

Remove contents. Damp wipe surfaces using a suitable cloth impregnated with an appropriate cleaning agent. Dry or allow to dry and replace contents.

Clean light diffusers

Ensure that the light fitting is switched off and necessary safety precautions are followed. Remove diffuser. Wipe fitting, lamp and diffuser using a suitable cloth impregnated with an appropriate cleaning agent to remove dust and dirt. Dry with a lint free cloth. Replace diffuser.

Clean microwave oven

Remove internal grease spillages and food debris using a cloth impregnated with a solvent free neutral detergent. Damp wipe external surfaces using a neutral detergent solution. Polish dry with a lint free cloth to remove cleaning marks.

Clean mirrors

Wipe using a suitable cloth impregnated with an appropriate cleaning agent to remove dust, dirt and superficial marks. Polish dry with a lint free cloth to remove cleaning marks.

Clean refrigerators

Disconnect from mains supply and remove contents. Damp wipe internal surfaces and shelves using a suitable cloth wetted in a solution of sodium bicarbonate. Replace contents and reconnect supply. Wipe external surfaces with a suitable cloth impregnated with a proprietary cleaning chemical. If necessary, polish dry with a lint free cloth to remove cleaning marks.

Clean stainless steel sinks/drinking fountains

Remove stains and residues inside the sink/drinking fountain using a suitable cloth impregnated with a proprietary stainless steel cleaner. Rinse with water then wipe using a suitable cloth. Polish dry to remove cleaning marks. Wipe external surfaces, fittings and traps using a suitable cloth impregnated with an appropriate cleaning agent. Polish dry to remove cleaning marks.

Clean toilets and urinals/sluices

Clean the internal surfaces of toilets and urinals/sluices using appropriate equipment and a proprietary toilet cleaner. Wipe external surface of fittings, taps, traps, visible cisterns and pipes using a suitable cloth impregnated with a neutral detergent solution. Ensure that all hinges, channels and outlets are cleaned. Remove debris from drain grilles.

Clean televisions

Remove dust, dirt and superficial marks from screens using a soft dry lint free cloth and a proprietary antistatic cleaning agent. Damp wipe casing using a suitable cloth impregnated with a neutral detergent taking care that necessary safety precautions are followed. Polish dry with a lint free cloth to remove cleaning marks.

Clean washbasins, showers and baths

Remove stains and residues inside the washbasin/shower/bath using a suitable cloth impregnated with an appropriate cleaning agent. Ensure that the splashback is also cleaned. Rinse with water then wipe using a suitable cloth. Wipe external surfaces, fittings and traps using a suitable cloth impregnated with an appropriate cleaning agent.

Clean (wooden) skirtings

Remove dirt and other contamination from skirtings using a suitable cloth impregnated with an appropriate cleaning agent. If necessary remove stubborn marks using a scrubbing brush. Wipe dry with a lint free cloth to remove cleaning marks.

Damp dust
Damp wipe

Using a suitable cloth moistened with a hand sprayer containing a neutral detergent solution remove dust, dirt and superficial marks from all surfaces. Polish dry with a lint free cloth to remove cleaning marks.

Damp mop

Remove dirt, soil and spillages from the floor using suitable mopping equipment and a solution of neutral detergent which is replenished at regular intervals.

Damp wipe and dry marble walls/tiled walls/pillars/partitions/panels/laminates

Remove dust, dirt and superficial marks using suitable equipment and an appropriate cleaning agent from the surface being cleaned. If necessary use an abrasive pad to remove stubborn marks. Polish dry with a lint free cloth to remove cleaning marks.

Damp wipe wastebins

Remove ingrained and superficial soiling from all surfaces of the bin using an abrasive pad or scrubbing brush with the assistance of a neutral detergent solution. Rinse and wipe dry with a lint free cloth.

Deck scrub

Remove ingrained and superficial soiling from the floor using a deck scrubbing brush and neutral detergent solution or other appropriate cleaning agent. Rinse by damp mopping with clean water.

Descale sanitary fittings

Clean internal surfaces using appropriate equipment and a proprietary toilet cleaner. Apply a proprietary acid descaler using a sponge applicator or by some other suitable means and allow to remain in contact with the surfaces for several minutes to allow chemical reaction to occur. Scrub using a stiff brush and flush thoroughly with copious amounts of water. Repeat the process if necessary until all traces of metal salts, uric salts or limescale are removed.

Dry dust

Using an appropriate dust control tool or lint free cloth, remove dust, fluff debris and dust particles from all surfaces.

Dust control mop

Remove dust, fluff and debris from hard floor areas using a suitable dust control mop in such a way that the mop is swept with a continuous stroke without being lifted from the surface.

Dust door/wall vents

Remove dust, fluff and debris from air vents using a suitable dust control tool. Remove stubborn marks with a suitable cloth impregnated with an appropriate cleaning agent. Polish dry with a lint free cloth to remove cleaning marks.

Dust walls/panelling

Remove dust and fluff from walls/panelling, including high ledges using an appropriate dust control tool or vacuum cleaner with suitable attachments.

Empty and damp wipe ashtrays/ashbins

Empty the contents of the ashtray/ashbin into a metal container and wipe with a suitable cloth impregnated with an appropriate cleaning agent. Polish dry any metalwork with a lint free cloth to remove cleaning marks.

Empty wastebins and dispose of rubbish

Empty waste into refuse bags and convey collected waste to designated disposal points. Replace bin liners in wastebin as necessary.

Full vacuum/vacuum clean

Remove dust and debris from the entire floor area using a vacuum cleaner.

High dust

Remove dust and fluff from horizontal and vertical surfaces, ledges and fittings above 2m high using suitable dust control apparatus or a vacuum cleaner with appropriate extension tubes.

Hose

Remove ingrained and superficial soiling using a pressure washer supported if necessary by a brush and an appropriate cleaning agent.

Lift and beat coir matting

Lift mat from well and transfer to a suitable external area where beating may be carried out. Place mat face down and beat with a suitable device to remove trapped particles of dust. Vacuum and then damp mop the mat well allowing it to dry before returning the mat to its normal position. Vacuum clean the mat once in place.

Machine scrub and dry

Remove ingrained and superficial soiling from the floor using a scrubber dryer fitted with suitable brushes and charged with a neutral detergent.

Mechanically sweep

Remove all loose dirt and debris from the floor using a motorised sweeper.

Neutral scrub

Loosen ingrained and superficial soiling on the floor using a suitable scrubbing machine fitted with appropriate brushes or pads and charged with a neutral detergent. Remove water and superficial soil from the floor by means of a mop regularly rinsed in clean water or by means of a wet pick-up device.

Neutral scrub and burnish

Loosen ingrained and superficial soiling from the floor using a suitable scrubbing machine fitted with appropriate brushes or pads and charged with a neutral detergent. Remove water and superficial soil from the floor by means of a mop regularly rinsed in clean water or by means of a wet pick-up device.

Once dry burnish the floor to achieve a shine free from scuff, abrasion and heel marks using a high speed or ultra high speed machine fitted with a suitable brush or pad.

Neutral scrub and polish

Loosen ingrained and superficial soiling from the floor using a suitable scrubbing machine fitted with appropriate brushes or pads and charged with a neutral detergent. Remove water and superficial soil from the floor by means of a mop regularly rinsed in clean water or by means of a wet pick-up device.

Once dry apply a light coat of polish to the floor using a suitable application technique and burnish using a rotary machine fitted with an appropriate floor pad to produce a shine.

Polish tops of desks and tables

Using a suitable lint free cloth and proprietary spray polish remove dust, dirt and superficial marks and buff to achieve a shine.

Polish wood furniture

Using a suitable lint free cloth and proprietary wood furniture polish remove dust, dirt and superficial marks and buff to achieve a shine.

Remove chewing gum

Remove chewing gum from surfaces using a proprietary chewing gum remover and a technique appropriate to the choice of product taking care not to spread the contamination or damage the surface being cleaned.

Remove debris from upholstered furniture

Remove fluff and other debris from upholstered furniture using a suitable brush, or vacuum cleaner with upholstery attachment, taking care not to overlook gaps around the edges of cushions.

Remove graffiti

Remove graffiti with a proprietary graffiti remover chosen as appropriate to the type of surface being cleaned, using a suitable brush or abrasive pad and taking care not to damage the surface. Rinse with water.

Remove/pick up litter

Clear all surfaces, floors, paved areas etc. of any rubbish, litter, cigarette ends etc. Collect waste in refuse bags and convey to designated disposal points.

Remove spots and stains from carpets

Wet stains: Blot with uncoloured absorbent paper or cloths until no more staining material is removed. If the carpet remains discoloured proceed as for dry stains. If the carpet is not discoloured but the staining is sticky flush out with a suitable spray extraction machine charged with a proprietary carpet cleaning chemical.

Dry stains: Treat the stain with a proprietary stain removal chemical appropriate to the type of stain being removed having first checked an inconspicuous area to ensure that the product has no deleterious effect on the carpet. Blot with uncoloured absorbent paper or cloths until no more staining material is removed. If necessary flush out with a suitable spray extraction machine charged with a proprietary carpet cleaning chemical.

Replenish consumables

Restock spent consumable items such as liquid soap, toilet rolls, paper towels etc., as necessary.

Sand and reseal wood floors

Remove soil and existing worn floor seal from wood floors using a suitable sanding machine with vacuum attachment and skirt and fitted with an appropriate grade of sanding disc. Apply sufficient coats of a proprietary wood seal to present a satisfactory appearance and hard wearing finish.

Sand and reseal wooden stairs

Remove soil and existing worn floor seal from treads and risers using coarse glass paper and if available, a suitable sanding machine. Apply sufficient coats of a proprietary wood seal to present a satisfactory appearance and hard wearing finish.

Sanitise telephones

Wipe the body and handset using either a suitable cloth impregnated with a disinfectant solution or with proprietary sanitising wipes. Ensure that the cord and around the keys are not overlooked. Damp wipe coin box and internal and external surfaces of hood (if applicable).

Scrub stairs and stair nosings

Remove ingrained and surface soiling from stair treads, risers and nosings using an abrasive pad or scrubbing brush and an appropriate cleaning agent. Dry with a cloth or mop.

Scrub vinyl chairs

Remove ingrained and surface soiling using an abrasive pad or suitable cloth and an appropriate cleaning agent. Wipe with a dry cloth or paper towel to remove moisture and leave free from cleaning marks.

Scrub wastebins

Remove ingrained and surface soiling using an abrasive pad or suitable brush and an appropriate cleaning agent. Wipe dry.

Spot clean upholstered furniture

Wet stains: Blot with uncoloured absorbent paper or cloths until no more staining material is removed. If the fabric remains discoloured or sticky proceed as for dry stains.

Dry stains: Treat the stain with a proprietary stain removal chemical appropriate to the type of stain being removed, applying the chemical from a clean white absorbent towel, having first checked an inconspicuous area to ensure that the product has no deleterious effect on the fabric. Blot with uncoloured absorbent paper or cloths until no more staining is removed. Specialist upholstery cleaning may then be required.

Spot clean glass

Remove obvious marks from glass surfaces using a suitable cloth and an appropriate cleaning agent. Polish dry with a lint free cloth to remove cleaning marks.

Spot mop

Remove spillages or other obvious marks from the floor using suitable mopping equipment and an appropriate cleaning agent to restore the floor to an evenly clean appearance.

Spot vacuum

Remove obvious litter from the floor using a vacuum cleaner.

Spot wipe doors

Remove obvious marks from doors, door handles, pushplates and kickplates using a suitable cloth and appropriate cleaning agent. Polish dry with a lint free cloth to remove cleaning marks.

Spot wipe/spot clean

Remove obvious marks from walls, paintwork, partitions, ceilings, metal window frames etc. using a suitable cloth and an appropriate cleaning agent.

Spray clean

Dust control mop the floor. Remove ingrained and surface soiling from smooth floors using a suitable floor maintenance machine fitted with the correct cleaning pads and using an appropriate cleaning agent dispensed from a hand trigger spray.

Spray clean with suction

Remove ingrained and surface soiling from smooth floors using a suitable floor maintenance machine with vacuum attachment fitted with the correct cleaning pads and using an appropriate cleaning agent dispensed from a hand trigger spray.

Spray extract carpet

Vacuum clean the carpet and remove stains. (q.v.) Pre-spray using a proprietary carpet cleaning pre-spray. Clean using suitable spray extraction apparatus charged with a proprietary spray extraction product. Ventilate the area to facilitate drying.

Spray extract upholstered furniture

Vacuum clean the furniture and remove stains. (q.v.) Pre-spray using a proprietary upholstery cleaning product. Clean using suitable spray extraction apparatus fitted with an upholstery tool and charged with a proprietary upholstery cleaning chemical.

Spray strip and redress floors

Strip polish from floor by the application of a fine mist of a suitable detergent and use of a high speed rotary machine fitted with a spray stripping pad, vacuum attachment and skirt. When the floor is clean and dry apply sufficient coats of an appropriate polish, with intermediate buffing and sweeping between coats to present a satisfactory appearance with hard wearing characteristics.

Sweep

Remove all dirt, fluff, debris and refuse from the floor using a sweeping brush.

Traffic vacuum

Remove dust and debris from traffic lanes only using a vacuum cleaner.

Turn coir matting

Lift coir mat from mat well. Vacuum mat well. (q.v.) Turn mat laterally through 180° and replace in well.

Vacuum clean – see full vacuum

Vacuum clean ceilings/ceiling tiles

Remove dust and fluff from ceilings using a vacuum cleaner fitted with a suitable tool.

Vacuum clean edges and corners

Remove dust and debris from edges, corners and skirtings (where appropriate) using a vacuum cleaner fitted with an appropriate tool.

Vacuum clean upholstered furniture or screens

Remove dust, fluff and debris from all surfaces and crevices using a vacuum cleaner fitted with an appropriate tool.

Vacuum mat well – see Lift and beat coir matting

Vacuum traffic lanes – see Traffic vacuum

Vacuum under floor voids

Ensure that any smoke detectors beneath the void are deactivated. Vacuum to remove all dust and debris using a vacuum cleaner fitted with plastic nozzles and tubing.

Vitrify stone flooring

Strip existing finish from floor with a rotary scrubbing machine fitted with an appropriate pad and using a suitable stripping agent or detergent lubricant. When thoroughly cleaned apply a proprietary vitrification chemical or combination of chemicals and burnish with a steel wool pad or alternative abradant to develop a glass-like surface. Remove residual dust or metal filaments using a dust control mop.

Wash and dry tiled or painted walls

Use suitable wall cleaning equipment and an appropriate cleaning agent to remove superficial and ingrained marks. Wipe to leave surface free from cleaning marks.

* * *

Drafting of the above or similar definitions is the last stage in the preparation of a cleaning specification and the document is now complete. You are therefore ready for the next phase which is the preparation of the *Tender Return* documents. This is discussed in detail in Part Two.

SUMMARY

The last stage in the drafting of a cleaning specification is to present a series of definitions for each of the operations specified in the cleaning schedules. Some argue that precise definitions serve no purpose whilst others feel that they assist in monitoring the quality of service.

A number of definitions have been suggested. These may need to be complemented or modified to suit your own requirements.

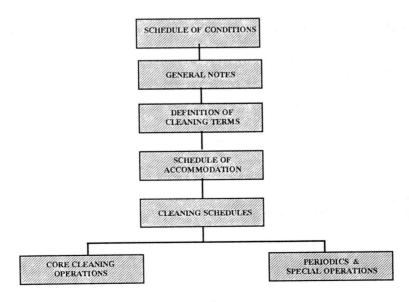

Part Two

Preparing the Tender Return

Chapter 9   Preparing the Tender Return Forms

9

PREPARING THE TENDER RETURN FORMS

You have now completed all of the stages involved in preparing a cleaning specification. If this has been done with the intention of managing your own labour you will have no need of a tender return document and should proceed directly to Part Three of this book.

If however the preparation of the cleaning specification was intended to be the first step in some form of competitive tendering exercise, you will now need to draft the **Tender Return**.

The *Tender Return* comprises a series of blank forms which the contractor is invited to complete. These forms are concerned with the following:

i.   The labour requirement
ii. The deployment of that labour
iii. The cost of labour
iv. Variation rates
v.   Equipment to be used
vi. Materials to be used
vii. The consumables to be provided
viii. The cost of periodics
ix. The cost of special operations
x.   Overhead costs
xi. The total contract price.

Let us consider each element in turn.

LABOUR REQUIREMENT

One of the first things the contractor will need to do in order to cost your specification is to calculate the manpower requirements. How this is done is discussed in greater detail in Part Three. Indeed it is suggested there that you too will need to do the same in order to satisfy yourself that the contractors' estimates are correct.

Provide a blank form to be completed along the lines of that shown in Figure 9.1. As you will see from the column headings a number of separate details are required. These are:

Grade of staff: These may include cleaning operatives, handymen, janitors, waste disposal operatives, drivers etc. although supervisory and management staff are not included but appear on *Form 2*. (Figure 9.2).

This information is necessary in order that you may satisfy yourself that the right type (i.e. appropriately trained) staff are to be employed.

Shift period: Different grades of staff may well be working at different times of the day. For example, although your core cleaning operations may be carried out between 1700 hrs and 2000 hrs you may have also specified that day cleaners are required. This column on the form therefore enables you to check that the contractor intends to provide cover at all of the times you need it.

Number of staff: This provides vital information. It is important for four reasons.

- it forms a basis for the calculation of the contract price

- it provides a check that the contractor is supplying sufficient labour to meet the requirements of the specification

- it supports any monitoring you have planned throughout the life of the contract enabling you to confirm that adequate manning levels are being maintained

- it may be important as a head count in the event of a fire

Number of hours: Also important for pricing purposes these columns additionally show manning levels to be provided at weekends when premium pay rates may be involved.

Total weekly hours: This simply sums the other columns. It is surprising how many mathematical errors can occur in this summation!

For clarity it is advisable, at the head of the form, to provide explanatory notes indicating to the contractor exactly what needs to be entered on the form. Roger Blackburn wrote:

Indicate separately all grades of staff to be employed including working supervisors but with the exception of non-working supervisory staff and management grades. Show the shift length, the numbers of staff employed on each shift, and the total number hours for each grade of staff.

If your specification embraces more than one building ask the contractor to provide the information on a building by building basis.

You will need a separate form to be completed to indicate the allocation of supervisory and management staff. Figure 9.2 shows a typical example. You will note that this form is identical to that illustrated in Figure 9.1 except that it additionally asks for clarification of supervisory arrangements for weekend working. Of course if no weekend working is envisaged this may be omitted.

Apart from the obvious reason of cost, you will be interested in the allocation of supervisory effort in order to ensure that the contractor intends to provide enough.

Furthermore, in future monitoring exercises you will want to confirm that adequate supervision is being maintained. You can only do this if you know what was budgeted from the onset.

Add an explanatory note at the head of the form along the lines of that shown above for *Form 1*. Again if you have more than one building ask for a building by building breakdown.

LABOUR DEPLOYMENT

Whilst *Forms 1* and *2* show the allocation of labour according to grade, you will also need to know something about its deployment. For example, some will be engaged in core cleaning operations, others will be carrying out periodics. Yet others may be involved in day cleaning, or waste management or external cleaning. An indication of labour deployment is especially important if more than one building is involved since you will want to know how the labour is distributed amongst the different buildings. A suitable form is illustrated by Figure 9.3. In this example two separate buildings are involved.

You will also note that in the East Building the total hours to be spent each week on 'Hi-Tech' cleaning has been broken out. This has been done in this instance to underline the importance attached to Hi-Tech cleaning in this particular building.

LABOUR COSTS

Form 4, illustrated by Figure 9.4 asks the contractor to indicate what rates he intends to pay his employees. Apart from the obvious reason that the hourly rate of pay affects the calculation of total contract price there are two other reasons why you may require this information. Firstly it shows that the contractor intends to pay a reasonable wage, thereby attracting a more permanent workforce with the attendant benefits that offers. Secondly by requiring an indication of wage levels at the beginning of the contract you can monitor that he is indeed paying those amounts throughout the life of the contract. As noted earlier it is not unknown for more unscrupulous contractors to declare a higher wage rate than they intend to pay thereby disguising an element of profit they plan to make at the expense of their workforce.

The form requires the rate for all grades of staff to be shown for both normal working and overtime working. An explanatory note at the head of the form might be:

Indicate below the rate of pay for each grade of staff to be employed on the contract including supervisory and management rates. Differentiate between weekday and weekend rates of pay.

On the same form you may also request details of wage rates to be paid for work carried out beyond the scope of the contract. You may need a special clean-up for a VIP visit for example. Or some emergency work – after a flood or fire perhaps, may be required. A suitable note might be:

Indicate rates of pay for each grade of staff should they be required to carry out work beyond the scope of the contract. Include supervisory and management rates and differentiate between weekday and weekend working.

VARIATION RATES

It may be, during the life of the contract, that portions of the building are changed. You may change a number of offices into an executive dining room for example, or install additional toilets to accommodate an increase in numbers of staff. Clearly such changes may well affect the total contract price. However it is far better to establish the effect of these variations from the outset than to haggle over costs at the time the alterations are made.

Using the *Schedule of Accommodation* and the cleaning schedules, the contractor will be able to calculate exactly what production rate he expects to achieve for each category of room within your classification of room types. Thus he will be able to calculate the cost over say 28 days of cleaning 10 square metres for each type of room.

Form 5 shown in Figure 9.5 asks for the costs for cleaning 10 square metres of each of the different types of accommodation over 28 days to be given. You will note that the form also asks for the rates that will apply in Year 2 of the contract.

Roger Blackburn's explanatory note read:

During the term of the contract, Neptune Stereophonics may vary the Schedule of Accommodation by making deletions, additions, or variations in the designated use of rooms. The contract price may vary accordingly. Indicate below the rates which apply for the cleaning of 10 sq. metres of each category of room over a period of 4 working weeks. Show the rates for each year of the contract.

You will also note that a box has been provided for the contractor to indicate the method by which he has calculated the rates.

EQUIPMENT

Once you have considered the labour involvement and deployment you must also scrutinise what equipment the contractor intends to use. Choice of equipment may well affect productivity and hence the total cost of labour. Labour costs are the biggest single component of the contract price. Furthermore you will want to confirm that he intends to use equipment that is best suited to the job. If for example you have many square metres of warehouse space to sweep and scrub, you will be looking to see a 'ride-on' type sweeper/scrubber with a high productivity rather than one (or several) pedestrian operated machines which demand a high labour involvement with associated poor productivity.

Form 6 shown in Figure 9.6 is adequate.

MATERIALS

There are several reasons why you will want to know what materials the contractor intends to use.

Firstly you will want to confirm that the materials chosen are not likely to have any deleterious effect on your floors, furniture, fixtures and fittings.

You will want to be satisfied that the chemicals are manufactured by a reputable manufacturer and are therefore likely to meet minimum standards of product uniformity and be accompanied by appropriate Safety Data Sheets.

You will want to be sure that the contractor intends to provide an adequate range of materials and that he does not include those chemicals which are not necessary. For example if you have a requirement in your cleaning schedules to *sanitise telephones* you will expect to see a telephone sanitiser on the list.

Finally, when you are subsequently monitoring the contract you will periodically want to check that the products in the cleaners' cupboards are those in the tender submission.

Form 7 illustrated by Figure 9.7 is suitable for these requirements.

CONSUMABLES

In the general notes you will have detailed the consumables which you expect the contractor to provide and given some indication of the quantities involved. Using this indication, the contractor will be able to quote a price for the provision of these consumables. This information is shown on *Form 8* (illustrated by Figure 9.8). It is important information because the consumables are additional to the contract price and by definition, are dependent upon the number consumed. You will therefore want to ensure that you are obtaining them at the best possible price. (During your monitoring exercises you will also want to ensure that you are paying only for those supplied).

PERIODICS

The need to cost periodics separately has already been discussed. In order to do so however it will be necessary for the contractor to show prices for each different task in the tender return. The appropriate form (*Form 9* in Figure 9.9) lists all of the tasks which appear in the *Periodic Schedule* (which in the example of Neptune Stereophonics coincides with the schedule shown in figure 5.6) but does not include special operations which are introduced later in the tender return document. You will also note in Figure 9.9 that a price for each single operation is required as well as a price for the total number scheduled over the full year. By requiring the information to be presented in this way you are able to consider the financial consequences of asking the contractor to undertake periodics at a greater frequency, whilst at the same time you can review the potential savings of a reduced frequency. At the bottom of the form are the words '*Transfer to Contract Price Form*'. The significance of this will be revealed in due course.

Draft a note at the head of the form as follows:

Indicate below the cost for the provision of periodic operations as detailed in the Periodic Schedule. Costs should include all overheads and profit.

SPECIAL OPERATIONS

In Chapter 5 we considered the reasons why certain operations might not be included in the *Periodic Schedule* even though they are carried out less frequently than monthly. These include tasks that may well be subcontracted; tasks that may be of sufficient importance to be considered separately; or tasks where it is particularly important to scrutinise the costs (such as window cleaning for example). These tasks have been described as *Special Operations*. Normally these will simply be included in the *Additions to Tender* section of the Tender Return (*Form 11*) which will be discussed shortly. Occasionally however you may wish to obtain a more comprehensive breakdown – if the task is particularly involved, as in the case of window cleaning, or carpet cleaning for example. *Form 10* illustrated by Figure 9.10 shows a typical example of a request for a more detailed breakdown of window cleaning costs. In those instances where you do demand this level of explanation, it is likely that you will have also written a detailed schedule for the operation as part of your general specification.

Tasks such as deep cleaning of kitchens, deep cleaning of toilets and Hi-Tech cleaning, seldom warrant a separate form like that shown in Figure 9.10 as *Form 10* but instead are included in the *Additions to Tender* section as we shall see.

ADDITIONS TO TENDER

Form 11 is concerned with additional costs not so far specified. These include the cost of the materials and equipment referred to in *Forms 6* and *7*; various administrative costs and overheads; the cost of special operations not already detailed; and profit. When scrutinising the tender return you will want to confirm that these costs coincide with demands made in your *General Notes* and statements made by the contractor. For example if a charge is to be levied for holiday and sickness costs, then you will want to be sure that the contractor pays his staff whilst they are on holiday or sick leave. This particular form also contains important information you will want to use when monitoring the contract. Let us examine *Form 11* in Figure 9.11 in greater detail.

Item 1. Equipment and Materials:

This entry will show the cost of items shown on *Forms 6* and *7*, in the case of the equipment amortised over the life of the contract, and will include the cost of maintenance of the equipment.

Item 2. Protective Clothing Costs:

Here the contractor is required to show his costs for protective clothing. By itemising the individual costs you have an opportunity to confirm that he intends to provide everything that you perceive as being necessary. For example if you have any hazardous wastes you will be looking to see that suitable gloves, overalls, masks etc. have been detailed; if outdoor working is necessary he should show provision of suitable outdoor wear. Look also to see that he has included laundry costs. If he has not, ask why not.

Item 3. Administrative Costs:

Here the contractor should show his general allowances for administrative overheads which will include head office support, employer's national insurance contributions, transport costs (if operatives are transported to the site), training costs etc. Different contractors will use different methods to calculate these training costs and you should therefore be sufficiently flexible to accommodate the different bases that are used but inflexible in demanding an adequate explanation as to how the costs are derived.

Item 4. Management Costs:

There are a number of reasons why you may want to separate management costs both from the cost of labour as calculated using the information shown on *Form 4* and from the administrative costs as shown in Item 3 above.

For example, you may want to clearly differentiate the value of the management input. There are two reasons why you may wish to do this. Firstly you may want to judge whether you can afford it. Secondly, and perhaps more importantly you may want to evaluate the cost-effectiveness of the management throughout the life of the contract for there can be no doubt that the success of a cleaning contract is highly dependent upon the quality of its management.

Additionally you may wish to see what involvement the contractor's 'head office' intend to have. Clearly you will expect frequent visits from them to satisfy yourself that they are monitoring the performance of their staff, and if you have specified in your *Schedule of Conditions* that monthly meetings are to take place between yourself and a senior management representative from the contractor, those meetings may well be costed under this item.

Item 5. Annual holiday/sickness costs

If the contractor offers paid leave to his staff the costs of that leave will be shown here. This presumes that the contractor intends to provide replacement labour perhaps from a pool or perhaps by way of overtime using existing staff, to ensure that you still receive the same daily input of effort. If instead tasks are omitted or frequencies are reduced because cleaning operatives are on holiday you will be looking for a *reduction* in the contract price.

The contractor should also have historical data to allow him to calculate what level of absenteeism he is anticipating due to sickness.

This might normally be absorbed into his general overheads but should be shown here instead. Make sure the amount being charged matches staff entitlement.

Item 6 et seq. Special Operations

The next series of items relates to those special operations where you require a separate detailed breakdown of costs in contrast to those shown in the periodic costings form (*Form 9*) or as shown as an example for window cleaning in *Form 10*. Note that each of these separate entries require the costs of overheads and profit to be included in the price as did the periodic costings. This makes it simpler to calculate the cost or savings arising from increasing or decreasing the frequency of such operations.

You will see in Items 6 and 7 on *Form 9* that three of each of these operations have been scheduled per annum. The contractors entry in this section will therefore indicate the price per operation as well as the price if the specified annual frequency is met. Item 8, *Hi-Tech cleaning* will only be included perhaps if your specification demanded specialist input for the cleaning of Hi-Tech equipment and was thus separated from your core cleaning schedules. You will note in Item 9 for carpet cleaning that both spray extraction which is a periodic technique and powder cleaning which is an interim technique, have been itemised separately. This gives you greater flexibility to vary your carpet cleaning routines. Some prefer as an alternative, to ask for carpet cleaning to be costed by the 100 sq. metres. However if this practice is adopted it is no longer in the format of a 'price per operation' and cannot therefore be easily incorporated into the total contract price as we shall see calculated on *Form 12*. Instead it will remain as an additional cost over and above the budgeted total and is somewhat inconsistent with the format we have otherwise adopted. If you want to specify carpet cleaning costs in greater detail consider drafting a form along the lines of that shown for window cleaning in *Form 10*.

All that is now required to conclude the preparation of *Form 11* is a suitable explanatory note at the head of the form. The following explanation will suffice.

Complete the form below giving whatever explanation is necessary to show how your figures have been derived. Make sure you include the costs for overheads and profit where directed to do so. Transfer the annual costs so calculated to the appropriate sections on Form 12.

TOTAL CONTRACT PRICE

Form 12 shown in Figure 9.12 brings together all of the calculations shown on *Forms 1 – 11*. Additionally it asks the contractor to specify what profit he is seeking. You will observe that the cost for each of two years is required in this example for Neptune Stereophonics. If the cost for each year is the same, be suspicious. It may be that the contractor has added the cost of year one to year two and divided by two. In this way you will be paying more than you need in year one with your money sitting in the contractor's bank account rather than your own.

Note also the following:

In line 3, *annual labour cost* involves multiplying the weekly cost by 52.143 which is the actual number of weeks in a year.

Profit in line 9 relates only to core cleaning tasks. The profit on periodics and special operations has already been included in the figures detailed on *Form 11*.

Line 18 invites the contractor to offer you a discount for prompt payment. This may be alluded to in your *General Notes*. However, if you have not specified a time period where prompt discount may apply in the *General Notes* add the words shown in italics in line 18 'for payment within n days' so that the contractor can indicate his own rules for discounting. If the contractor allows no discount lines 17 and 19 will be the same.

The *Tender Return* document is now almost complete. However a little additional work still needs to be done.

First you will need to draft a form for the contractor to complete confirming that he understands all of your requirements and agrees to supply a cleaning service for an agreed cost. A typical example is shown in Figure 9.13.

Next you should prepare a *non-collusive tendering certificate*. This is the document whereby the contractor agrees not to fix prices in collusion with any other party. An example is shown in Figure 9.14.

Finally you may wish to prepare a check list as an *aide memoire* to the contractor in order to ensure that he provides all of the supporting information that you require. Roger Blackburn drafted the following to precede the forms detailed in Figures 9.1 to 9.12:

The following check list shows the supplementary information you should supply to support your tender.

1. *An overview of the proposed modus operandi.*

2. *Details of your employment terms and conditions; recruitment policy; company management structure; annual leave entitlement and arrangements for providing holiday and sickness cover.*

3. *Your policy for training staff of all grades indicating methods, frequency and facilities for training that are available.*

4. *Your procedures for implementing quality control indicating methods, frequency and follow-up in the case of failures being identified.*

5. *Health and Safety policy including methods for conformance with COSHH, Electricity at Work and Lifting and Handling regulations.*

6. *Name, addresses and telephone numbers of at least three other similar clients*

of your company to which reference may be made.

7. *Full details of your insurance cover including the names of your insurers, relevant policy numbers and the amount of the cover.*

8. *Audited annual accounts for the last two financial years.*

9. *Bank details.*

Additionally you should complete forms 1 to 12 and sign the two declarations at the end of this document.

The *Tender Return* document is now complete and you are ready to go out to tender.

* * *

In Chapter 7 some aspects of the timescale were considered and a typical timetable was suggested. Hopefully, by the time your *Specification* and *Tender Return* document have been completed you are ready to decide which of those who have responded to your advertisement are to be invited to proceed to the next stage. If you did not place an advertisement then you should have already written to a number of contractors asking if they would like to quote.

Six is an ideal number for a large site. Try to invite them all to come on the same day and at the same time for the reasons given in Chapter 7 (economy of time; fairness; equal dissemination of information; etc.).

In your letter of invitation to tender you will also need to inform the contractors of your timetable. It should contain the following information:–

i. The date and time of the site visit.

ii. The last date and time for submission of their quote.

iii. The date by which they will be notified whether or not they have been short-listed for interview.

iv. The date of the interview for short-listed contractors.

v. The date on which they will be notified of your final decision.

vi. The date on which the contract is to begin.

Some organisations, especially in the public sector, will have their own strict rules for invitations to tender. Indeed in the public sector these will need to satisfy EU requirements. The procedure detailed above may however be considered as a reasonable guideline for those who do not need to fulfil some mandatory procedure. If in doubt the overriding principle should be one of fairness.

During the interval, whilst the contractors are preparing their submission, you may wish to be making your own estimations of manpower requirements. If this is your intention a suitable technique is discussed in Part 3.

SUMMARY

The Tender Return document ensures that each contractor provides you with information in exactly the same manner. This facilitates comparison thereby allowing a more objective decision to be taken.

In all there are twelve forms to be completed. These are:

- Forms for the allocation of labour, supervision and management.

- Forms showing how the labour is deployed.

- Forms analysing the cost of the labour.

- A form to show the relative cleaning costs for different types of accommodation.

- Forms detailing equipment, materials and consumables.

- Forms showing periodic costs and the cost of special operations.

- A form detailing additional costs including overheads and profit and another summarising the total contract price.

Figure 9.1: Form showing allocation of manual labour

NEPTUNE STEREOPHONICS

FORM 1. LABOUR ALLOCATION – All grades except Non-Working Supervisory and Management staff

Grade of Staff	Shift Period	No. of Staff	Number of Hours							Total Weekly Hours
			Mon	Tue	Wed	Thur	Fri	Sat	Sun	

Figure 9.2: Form showing allocation of supervisory and management effort

Grade of Staff	Shift Period	No. of Staff	Number of Hours							Total Weekly Hours
			Mon	Tue	Wed	Thur	Fri	Sat	Sun	

NEPTUNE STEREOPHONICS

FORM 2. LABOUR ALLOCATION – Supervisory and Management Staff

Please state how weekend work will be supervised

Figure 9.3: Form showing deployment of labour

Grade of Staff	Shift Period	No. of Staff	Number of Hours							Total Weekly Hours
			Mon	Tue	Wed	Thur	Fri	Sat	Sun	
NEPTUNE STEREOPHONICS — **FORM 3. DEPLOYMENT OF LABOUR**											
NORTH BUILDING										
Core Cleaning 0500–0800 hrs										
External Areas										
Periodic Cleaning										
Supervision										
TOTAL FOR EAST BUILDING										
EAST BUILDING										
Core Cleaning 1700–2000 hrs										
Periodic Cleaning										
Hi–Tech Cleaning										
Supervision										
TOTAL FOR EAST BUILDING										

Figure 9.4: Form showing cost of labour

NEPTUNE STEREOPHONICS

FORM 4a. LABOUR COST ANALYSIS

Grade of Staff	Rates per Hour (£)		
	Mon–Fri	Sat	Sun

Form 4b. *EXTRA* CONTRACT LABOUR COSTS

Grade of Staff	Rates per Hour (£)		
	Mon–Fri	Sat	Sun

Figure 9.5: Form showing schedule of rates for variations to the Schedule of Accommodation

NEPTUNE STEREOPHONICS

FORM 5. SCHEDULE OF RATES

Room Category	Year 1 Rate for 4 weeks per 10 sq.m.	Year 2 Rate for 4 weeks per 10 sq.m.
EN		
OF		
CI		
TO		
RE		
CO		
PA		
WA		
ST		
SR		
SC		
ME		
KI		
PR		
SP		
EX		

Please indicate how these rates have been calculated

Figure 9.6: Form showing equipment to be used

NEPTUNE STEREOPHONICS FORM 6. EQUIPMENT TO BE USED			
Purpose	Manufacturer	Model Details	Number of Machines

Figure 9.7: Form showing materials to be used

NEPTUNE STEREOPHONICS		
FORM 7. MATERIALS TO BE USED		
Product Name	Manufacturer	Purpose

Figure 9.8: Form detailing consumables to be provided

NEPTUNE STEREOPHONICS FORM 8. CONSUMABLES			
Item	Manufacturer and Brand Name	Quantity per Box/Pack	Unit Price (£)

Figure 9.9: Form showing periodic costings

NEPTUNE STEREOPHONICS FORM 9. PERIODIC COSTINGS			
Task	Price Per Operation £	No of Operations per Annum	Total Annual Cost £
Sweep Floors		4	
Neutral scrub vinyl floors		4	
Damp wipe partitioning between offices		4	
Wash and dry goods lift walls		4	
Spot clean and vacuum clean upholstered screens		6	
Vacuum clean upholstered chairs		6 4 2	
Hose brush factory doors		2	
High dust high racking		2	
Remove, dry clean and rehang curtains		1	
Vacuum vertical blinds		4 2	
Damp wipe venetian blinds		2	
Clean diffusers and change light tubes		1	
Remove algae from roof		1	
Clean overhead pipework and ducting		1	
TOTAL ANNUAL COST (Transfer to Contract Price Form)			

Figure 9.10: Form showing breakdown of window cleaning costs

NEPTUNE STEREOPHONICS **FORM 10. WINDOW CLEANING COSTINGS**			
Task	Price Per Operation £	No of Operations per Annum	Total Annual Cost £
Clean both sides of internal glass partitions and panels		4	
Clean internal and external glass of all windows in North building		8	
Clean internal and external glass of all windows in East building		12	
Clean internal and external glass partitions, panels and windows to Main Reception		26	
TOTAL ANNUAL COST **(Transfer to Contract Price Form)**			

Figure 9.11: Form detailing Additions to Tender

NEPTUNE STEREOPHONICS
FORM 11: CALCULATIONS OF ADDITIONS TO TENDER

£

1. **Equipment & Materials**
   Equipment
   Maintenance of Equipment
   Materials
   Total Annual Cost
   _____

2. **Protective Clothing Cost** (specify)

   Total Annual Cost
   _____

3. **Administrative Costs**
   General overhead
   Staff NI
   Others (please specify)
   Total Annual Cost
   _____

4. **Management Costs** (specify)

   Total Annual Cost
   _____

5. **Annual Holiday/Sickness Costs** (specify)

   Total Annual Cost
   _____

Figure 9.11: Form detailing Additions to Tender

NEPTUNE STEREOPHONICS

FORM 11: CALCULATIONS OF ADDITIONS TO TENDER CONT'D

£

6. **Deep Cleaning of Kitchens** (x 3 operations)

   Cost per operation (including overheads & profit) _____

   Total Annual Cost _____

7. **Deep Cleaning of Toilets** (x 3 operations)

   Cost per operation (including overheads & profit) _____

   Total Annual Cost _____

8. **Hi-Tech Cleaning**

   (including overheads & profit) _____

   Total Annual Cost _____

9. **Carpet Cleaning**

   Spray Extraction (x 3 operations)
   Cost per operation (including overheads & profit)

   Dry Powder Cleaning (x 4 operations)
   Cost per operation (including overheads & profit)

   _____

   Total Annual Cost for Carpet Cleaning _____

Figure 9.12: Form showing calculation of total contract price

NEPTUNE STEREOPHONICS FORM 12. CONTRACT PRICE	Weekly Hours	Year One £	Year Two £

1. Weekly Labour Hours and Cost
    – Routine Cleaning
    – Day Cleaning
    – Supervision			

2. Sub Total: Weekly Labour Cost			
3. **Annual Labour Cost (Line 2 x 52.143)**			
4. Equipment & Material Costs			
5. Protective Clothing Costs			
6. Administrative Costs			
7. Holiday/Sickness Costs			
8. Management Costs			
9. Profit			
10. Sub total: Net Annual Cost (Σ Lines 3–9)			
11. Periodic Cleaning Costs			
12. Window Cleaning Costs			
13. Kitchen Cleaning Costs			
14. Toilet Cleaning Costs			
15. Hi-Tech Cleaning Costs			
16. Carpet Cleaning Costs			
17. TOTAL CONTRACT PRICE (Σ lines 10–16)			
18. Prompt Payment Discount for *payment within n days*			
19. DISCOUNTED CONTRACT PRICE			

Figure 9.13 Example of Form of tender for the provision of cleaning services

NEPTUNE STEREOPHONICS

FORM OF TENDER FOR THE PROVISION OF CLEANING SERVICES

To: Neptune Stereophonics plc.
157–173 Prince Consort Road
Chorlton–cum–Hardy
Manchester M21 OAB

We have examined the Schedule of Conditions and the specification for the cleaning of the above premises and having satisfied ourselves that we fully understand the extent and the description of the accommodation to be cleaned and the standards that are required, we offer to provide this service in accordance with the Schedule of Conditions and Specification for the sums shown below.

We undertake to provide the complete service as set out in the above mentioned documents, for a sum of £ .... per annum, (amount in words) .... in the first year of the contract and for the sum of £ .... per annum, (amount in words) .... in the second year.

These figures are given exclusive of VAT.

This tender, together with your written acceptance thereof, shall constitute a binding contract between us.

We understand that you are not bound to accept the lowest or any tender you may receive.

Signed: ....

Name in Capitals: ....

Position: ....

Duly authorised to sign for and on behalf of:

(In Capitals): ....

Address: ....

Tel: .... Fax: ....

Date:: ....

Figure 9.14 Example of Form of non-collusive tendering certificate

NEPTUNE STEREOPHONICS

NON-COLLUSIVE TENDERING CERTIFICATE

We certify that this is a bona fide tender, and that we have not fixed or adjusted the amount of the tender by or under or in accordance with any agreement or arrangement with any other person. We also certify that we have not done and we undertake that we will not do at any time before the hour and date specified for the return of this tender any of the following acts:-

    a. communicate to a person other than the person calling for those tenders the amount or approximate amount of the proposed tender, except where the disclosure, in confidence, of the approximate amount of the tender was necessary to obtain insurance premium quotations required for the preparation of the tender;

    b. enter into any agreement or arrangement with any other person that he shall refrain from tendering or as to the amount of any tender to be submitted;

    c. offer or pay or give or agree to pay or give any sum of money or valuable consideration directly or indirectly to any person for doing or having done or causing or having caused to be done in relation to any other tender or proposed tender for the said work any act or thing of the sort described above.

In this certificate, the word 'person' includes any person and any body or association, corporate or incorporate; and 'any agreement or arrangement' includes any such transaction, formal or informal, and whether legally binding or not.

Signed: ....

Name in Capitals: ....

Position: ....

Duly authorised to sign for and on behalf of:

(In Capitals): ....

Address: ....

Tel: .... Fax: ....

Date:: ....

Part Three

Estimating the Manpower Requirements

Chapter 10   Calculating the Manpower Requirements

Chapter 11   Deploying the Workforce

Chapter 12   Costing the Operation

10

CALCULATING THE MANPOWER REQUIREMENTS

Roger Blackburn was satisfied that he had now produced a professional cleaning specification which would serve his purposes, perhaps with occasional fine tuning, over several years. He had invited six contractors to quote to clean Neptune Stereophonics, had issued each with a full copy of the specification and had drafted a standard set of Tender Return forms on which they could quote on a uniform basis. His only concern now was how to evaluate the various quotes he was due to receive in three weeks time. Suppose for example, one quoted half the price of another. Would that be the right one to choose? Or would it mean that they had underestimated the requirement and would be unable to meet his objectives? But if he chose a more expensive option, how could he convince his own boss he was not wasting the company's money? There was only one way to be sure. He would have to carry out a workloading exercise.

Unfortunately the hard work is not yet over. If you intend to estimate the manpower requirements in order to check the contractors' calculations it is necessary to do some sums of your own. This means you will need to refer back to your *Schedule of Accommodation* and the detail you have collected in relation to room areas. The estimation of manpower is dependent upon the use of 'standard times' developed using well documented work study techniques. For many tasks the times are based upon area – hence the reason for recording the areas of the rooms in the *Schedule of Accommodation* as illustrated by Figure 3.1. Thus if we have a carpeted area measuring 100 square metres that requires a full vacuum every day from Monday to Friday and the standard time for vacuuming 100 square metres is 0.14 hours, then we know that 0.14 x 5 is the number of hours (0.7) required each week to vacuum that area i.e. 0.7 x 52.143 or 36.50 hours per annum. If labour (with overheads) costs £5.00 per hour then the annual cost of a daily vacuum of 100 sq. metres is 36.50 x 5 or £182.50. Perhaps instead you will only wish to full vacuum on a weekly basis! (This need to get the frequency right was also discussed in Chapter 4).

However, there are some tasks that are not costed by the 100 square metres but depend upon a standard time based upon the number of items to be cleaned. For example you may not wish to cost the cleaning of a wc. according to the number of square metres of toilet floor but according to the time taken to clean one w.c. For this reason there are some standard times that are based upon the number of units to be cleaned and not upon the floor area. Figure 10.1 lists tasks that are timed by floor area and Figure 10.2 lists those that are timed as 'unit counts'. Typical standard times are also shown for a 10,000 square metre building.

You will note nevertheless that certain cleaning tasks appear in both Figure 10.1 and 10.2. This is because some contractors prefer to cost by floor area, irrespective of the number of items involved, whilst others prefer to use unit counts. So, in Figure 10.1 the time to clean 100 sq. metres of toilet for example, is shown under *Clean toilet* as 1.00 hour. This includes cleaning all toilet fixtures, cleaning mirrors, wiping all surfaces, refilling dispensers, emptying waste bins and damp mopping the floor. However the times to clean a single wc, to clean the cubicle, to clean wash-basins etc. are shown as unit counts in Figure 10.2. The unit count technique is judged to be more accurate but may be too sophisticated for your needs.

Some items are always timed by the unit however. This means that in conducting your site survey for estimation purposes it is necessary to count certain objects and make a record of the total number. Clearly in the interests of maximum cost effectiveness you will do this at the same time that you carry out your original building survey when preparing your *Schedule of Accommodation* and *Cleaning Schedules*. So, although the Schedule of Accommodation given in your specification document may appear as shown in Figure 3.1 you may wish to draft *pro forma* blank sheets as shown in Figure 10.3 before commencing your initial site survey. As you will see the *pro forma* contains additional columns for you to make a tally of those items which require a unit count, and your working document will look something like that shown in Figure 10.4.

Different organisations use different values for standard times and the ones shown in Figures 10.1 and 10.2 are a guideline based upon pooled estimates. That is they are mean values from a number of independent sources.

It may well be that you prefer to develop your own set of times. This is probably the best way to arrive at suitable values since then you will have data that relates only to your own site. However the derivation of accurate work study data requires specialist expertise and hours of observation and is therefore likely to be infeasible.

Before we attempt an example calculation using the data given in Figure 10.1 and 10.2 there are two other aspects to consider. These are *building size* and *level of occupancy*.

Building size

If your building is only small a significant proportion of the time involved will be concerned with setting up and putting away cleaning equipment. If on the other hand, the building is very large, a smaller proportion of the time will be spent on these activities. To compensate, the following adjustments should be made to the times shown in Figure 10.1 which apply to a building of 10,000 square metres.

i) If the building is less than 2,500 square metres, *increase* standard times by 10%.

Thus, if the standard time for mopping 100 square metres as given in Figure 10.1 is 0.20 hours, then for a building of 2,500 square metres or less a standard time of 0.20 x 1.10 or 0.22 hours should be used.

ii) If the building is more than 2,500 square metres but less than 7,500 square metres, *increase* standard times by 5%.

Thus, using the same basic standard time for mopping 100 square metres from Figure 10.1 of 0.20 hours, the new standard time becomes 0.20 x 1.05 or 0.21 hours.

iii) If the building is between 7,500 and 12,500 square metres use the standard times shown in Figure 10.1.

iv) If the building is more than 12,500 square metres but less than 17,500 square metres, *decrease* standard times by 5%.

In our example then, the time for mopping 100 square metres in such a building becomes 0.20 x 0.95 or 0.19 hours.

v) If the building is more than 17,500 square metres *decrease* the standard time by 10%.

The time for mopping 100 square metres becomes 0.20 x 0.90 or 0.18 hours.

Level of occupancy

Level of occupancy refers not only to the number of persons using the premises but also to the density of furniture, fixtures and fittings. The two tend to be interrelated.

Clearly the greater the density or level of occupancy the lower is the cleaning efficiency. For example if a room measuring 100 square metres contains only one desk, chair, telephone and waste paper bin the time needed to clean that room will be significantly less than if the room has 20 occupants each with their own desk, chair, telephone and bin.

The values shown in Figure 10.1 should be adjusted as follows:

i) If the level of occupancy is judged to be low *decrease* the standard time by 20%.

Using the same example of mopping, the standard time for 100 square metres becomes 0.20 x 0.80 = 0.16 hours.

ii) If the level of occupancy is judged to be medium, use the standard time shown in Figure 10.1.

iii) If the level of occupancy is judged to be high *increase* the standard time by 20%.

The time for mopping 100 square metres in this case becomes 0.20 x 1.2 = 0.24 hours.

It may be argued that in the above example that furniture cleaning times will increase as occupancy levels increase whilst floor cleaning times will decrease because there is less unoccupied floor to clean. This possibility is ignored in the above calculations because (a) the manipulation of the standard times would become too sophisticated and hence time consuming if hundreds of such judgements had to be made; and (b) it may be equally argued that large open spaces can be mopped, vacuumed or whatever far more speedily than cluttered floors with numerous edges, corners, nooks and crannies.

There is one final word of caution. In the case of those tasks in Figure 10.4 which are timed as a unit take care not to adjust those times to account for different levels of occupancy or different building sizes. The time taken to clean an entrance mat is the same whether the building is 1,000 square metres or 100,000 square metres and whether 20 people use the building or 2,000 use the building. (You may however need to increase frequencies of cleaning in this latter instance but this would have been done when preparing your cleaning schedules).

Roger Blackburn is now ready to begin to workload the Neptune Stereophonics site. His first requirement is to collect together data according to his room classification system.

Figure 3.4 illustrates typical categories you may have chosen. For each of these you will have drafted a separate cleaning schedule. Your first objective therefore will be to sum floor areas and all items that are timed per unit for each of your categories Figures 10.5 and 10.6 will make this clear.

Figure 10.5 shows the blank *pro forma* which you will need to draft as a worksheet. Figure 10.6 shows one such *pro forma* partially completed for the accommodation at Neptune Stereophonics. When complete, the areas listed in the first column should coincide with those listed on the *Schedule of Accommodation* as illustrated by Figure 3.1 and the total area of office floor should coincide with the total given in the Summary sheet (Figure 3.9). You will also note in Figure 10.6 that staircases have been marked with an asterisk. These are normally timed per flight using the unit count principle rather than by the 100 square metres (Note. In this context one flight is the distance between two floors). Lifts are also timed in the same way.

The next step is to calculate the manpower necessary to meet the requirements of the specification for each category of room within your classification. As an example, let us consider the manpower required to clean the office accommodation. A suitable *pro forma* to assist in the calculation is shown in Figure 10.7. This has been completed in Figure 10.8 using the specification for offices shown in Figure 4.1. Using the form shown in Figure 10.7 and the specification from Figure 4.1, Figure 10.8 should be competed as follows.

I. Enter the name of the building. Your site may accommodate more than one building. It is important not to get them confused otherwise you may finish up with the wrong labour deployment on a building–by–building basis.

II. Indicate the room classification. In the example shown in Figure 10.8 we are concerned only with carpeted offices. Offices with vinyl floors would be detailed separately. (For the purpose of this illustration let us assume every office is carpeted).

III. Enter the density of occupation – in this case *medium*.

IV. Enter the total area for the room category. The value of sq. metres is taken from Figure 10.6 (and coincides with the value given in Figure 3.9).

V. Enter all the core cleaning operations for carpeted offices as shown in Figure 4.1. Note that offices with vinyl floors would appear on a different manpower development sheet.

VI. Enter the total number of items timed as unit counts shown for offices in Figure 10.6. Figure 10.8 uses this data.

VII. Enter the standard times as given in Tables 10.1 and 10.2.

VIII. Calculate the hours per operation for all of the carpeted offices on site.

In the case of operations timed by the 100 sq. metres (i.e. those in Table 10.1) this value is (**IV** x **VII**) / 100. In the case of those timed as unit counts it is simply **IV** x **VII**.

IX. Note the annual frequency for each operation. Tasks listed in the specification as *daily* are carried out 260 times per annum for a five days per week cleaning operation. Those carried out monthly are carried out 12 times per annum. Figure 10.9 summarises annual frequencies for different service requirements.

X. Calculate the total hours spent on each operation during the year. This is obtained by calculating **VIII** x **IX**.

XI. Sum column **V** to find the total labour hours required to meet the requirements of the core cleaning schedule for all of the carpeted office accommodation on site. Let us call this value Σ_{OF}. In the example shown in Figure 10.8 Σ_{OF} = 2397.0 hours.

Once this is complete you will need to carry out the same exercise for every category within your room classification system.

Finally this will enable you to find the total time required in one year to meet all of the requirements of the core cleaning schedules. If this total annual time is expressed as:

$$\Sigma_a = \Sigma_{OF} + \Sigma_{EN} + \Sigma_{TO} + \Sigma_{CO} + ...\Sigma_n$$

Where Σ_{OF} is the total time to core clean office accommodation, Σ_{EN} is the total time to core clean entrances, Σ_{TO} is the total time to clean toilets, and so on in accordance with your classification system.

To find the daily requirement Σ_d simply divide the total annual requirement Σ_a by the number of service days. Thus if the site is cleaned 5 days each week the daily manpower requirement becomes:

$$\Sigma_d = \frac{\Sigma_a}{260} \text{hours}$$

Standard work study practice suggests that this value should be increased by allowances for relaxation, contingencies, and other special nonconforming activities such as set-up times, time spent getting from the cleaners store to the site where cleaning is to be carried out, (imagine the distances that may be involved in an airport for example), and changing time if special protective clothing is required. The value of Σ_d may therefore need to be increased by some percentage appropriate to the local conditions.

Once this exercise is complete you will know what manpower is necessary to meet the requirements of the core cleaning schedules. All that now remains is to determine what manpower is needed for the periodics and special operations.

The calculations are carried out in almost exactly the same fashion except that whereas before we were concerned with the total time required to carry out every task in a particular category of room, we are now concerned with the time to complete just one task and perhaps for the entire site.

For example, suppose it is specified that carpets are to be shampooed every three months; suppose also that there are 4,000 square metres of carpeted offices, 1,000 square metres of carpeted corridors and 100 square metres of carpeted reception areas. The calculation will be as follows:

Total area of carpet to be shampooed $= 4,000 + 1,000 + 100$

$= 5,100$ square metres

Time to rotary shampoo 100 square metres
(from Table 10.1) $= 1.25$ hours

Therefore time to shampoo 5,100 square metres $= \dfrac{5100 \times 1.25}{100}$

$= 63.75$ hours

Time to shampoo 5,100 square metres
4 times per annum $= 4 \times 63.75$

$= 255$ hours

Thus we know the *time per operation* to shampoo all of the carpets (and hence the cost) and also the *time per annum* to shampoo the carpets. As discussed in Chapter 9 this gives us the flexibility to change the carpet shampoo frequencies with a clear understanding of the consequence this change may have on cost.

Of course if the specification requires office carpets to be shampooed 6 monthly, corridor carpets to be shampooed 3 monthly and reception carpets to be shampooed monthly the basis for calculation becomes:

Office carpets: $\dfrac{4000 \times 1.25 \times 2}{100} = 100$ hours per annum

Corridor carpets: $\quad\dfrac{1{,}000}{100} \times 1.25 \times 4 = 50$ hours per annum

Reception carpets: $\quad\dfrac{100}{100} \times 1.25 \times 12 = 15$ hours per annum

Therefore total annual time requirement for carpet shampooing = 165 hours

In this fashion it is possible to calculate the time requirements and hence cost of each periodic activity.

There is one final useful piece of information which is obtained during the core cleaning calculations and that is the production rate which prevails for each of the different categories of room. This information is necessary to enable Form 5 of the Tender Return to be completed. *Form 5*, shown in Figure 9.5 asks for the cost of cleaning each category within your room classification system in case you decide to change the use of various rooms.

This information is derived from the data presented in Figure 10.7 by dividing the area IV by the sum of column VIII.

In the example for office accommodation illustrated by Figure 10.8 this is 3394/31.37 or 109.2 m²/hour. Thus to clean 10 square metres of **OFFICE** accommodation for 28 days (information requested in Figure 9.5) the cost will be:

$$(10 \times 28)/109.2 \times R$$

where R is the hourly wage rate.

Clearly a different production rate and hence a different cost per 10 square metres per 28 days will prevail for different categories of rooms, cleaned according to different schedules.

This is the type of explanation that should be inserted in the box at the bottom of Form 5.

Note: For simplicity the schedules used throughout this example assume that tasks scheduled *daily* are carried out every day. In fact such operations as *spot vacuum* or *spot mop* will only be carried out on those days when no *full vacuum* or *full mop* is carried out, i.e. in a 260 service day contract they will only be performed 208 times.

When completing calculations of the type described here this reduced frequency needs to be taken into account. In the worked examples shown however this has not been done to avoid unnecessary confusion.

SUMMARY

Using a series of standard times, it is possible to calculate the necessary manpower to meet the requirements of the cleaning specification. In the case of core cleaning operations these are best derived on a *room category by room category* basis according to the room classification system you have adopted.

In the case of periodics and special operations it is customary to calculate the manpower on an *operation by operation* basis. Such an approach allows greater flexibility in costing.

When applying standard times to the tasks defined in the cleaning schedules, it is important to remember than some operations are timed by unit area whilst others are timed by unit count. The data given in Figures 10.1 and 10.2 assume buildings of a particular size and density of occupation. Any variations from these 'average' calculations may necessitate an adjustment of the standard times.

Once the calculations are complete it is necessary to consider deployment of the workforce to achieve the calculated number of hours necessary to meet the requirements of the specification.

Figure 10.1: 'Standard' times for cleaning per 100 square metres of floor area in a typical medium occupancy building of 10,000 square metres.

Task	Time/100m² (hours)
FLOORS	
Apply anti stat/protector to carpet	0.13
Bonnet buff carpet	0.45
Buff floor (>600 rpm)	0.11
Buff floor (400–600 rpm)	0.21
Buff floor (<400 rpm)	0.31
Damp mop	0.23
Deck scrub	1.80
Dry powder clean carpet	1.25
Dust control mop	0.09
Full vacuum	0.14
Hose down floor	0.09
Machine scrub	0.70
Machine scrub and dry	0.90
Machine sweep	0.10
Pile lift carpet	0.40
Pressure wash	0.95
Redress with one coat	0.32
Redress with two coats	0.60
Remove chewing gum	0.02
Rotary shampoo carpet	1.25
Spot clean carpets	0.05
Spot mop	0.03
Spot vacuum	0.06
Spray clean (<400 rpm)	0.50
Spray clean (>400 rpm)	0.35
Spray clean with vac. attachment	0.25
Spray extract carpets	1.50
Strip	3.00
Sweep Floor	0.10
Remove stains and spillages	0.03
Vacuum edges and corners	0.03
Vacuum floor void	1.00
Vitrify	4.00
Wet mop	0.25
Wet pick up	0.18
PAINTWORK, WALLS, PARTITIONS, DOORS, CEILINGS AND GLASS	
Clean ceilings	4.00
Clean ceiling vents	0.13
Clean door brass	0.07
Clean door metalwork	0.03
Clean door pushplates and handles	0.03
Clean external windows	1.80
Clean glass in doors	0.03
Clean glass partitions	0.15
Clean internal windows	1.30
Damp wipe doors	0.06
Damp wipe walls	1.00
Dust light fittings	0.25
Dust walls	0.20
Dust walls over 2m	0.14
Spot clean glass in doors	0.01
Spot clean glass in partitions	0.06
Spot wipe doors	0.03
Spot wipe walls	0.03
Wash walls	3.50
Vacuum ceilings	1.70
Vacuum walls	0.85
FURNITURE, FIXTURES AND FITTINGS	
Clean air vents	0.07
Clean brasswork	0.07
Clean cupboard interior	0.14
Clean glass in furniture	0.03
Clean metalwork	0.03
Detail dust	0.20
Dust	0.03
Dust above 2m	0.10
Dust hi-tech equipment	0.05
Polish furniture	0.08
Spot clean chairs	0.03
Vacuum pipework	0.06
TOILETS	
Check clean toilet	0.50
Clean partitions of cubicles	0.45
Clean toilet	1.00
Replenish supplies	0.09
REFUSE	
Empty ash bins	0.03
Empty ashtrays	0.03
Empty wastebin	0.03
Empty wastebin and replace liner	0.04
MISCELLANEOUS	
Remove graffiti	0.02

Figure 10.2: 'Standard' times for cleaning individual items in a typical medium occupancy building.

Task	Time/Unit (hours)
FLOORS	
Clean entrance mat	0.02
Clean under entrance mat	0.05
PAINTWORK, WALLS, PARTITIONS, DOORS, CEILINGS AND GLASS	
Clean glass doors	0.03
Clean light diffusers	0.03
Clean mirror	0.01
FURNITURE, FIXTURES AND FITTINGS	
Change curtains	0.04
Clean drinking fountain	0.01
Clean glass fronted case	0.03
Clean light fitting	0.05
Clean public telephone	0.04
Clean sink	0.04
Clean TV monitor	0.03
Clean venetian blind	0.05
Clean vertical blind	0.05
Damp wipe telephone	0.02
Make bed	0.08
Polish top of furniture	0.03
Sanitise telephone	0.02
Scrub vinyl chairs	0.03
Shampoo upholstered chairs	0.10
Vacuum curtains	0.05
Vacuum upholstered chairs	0.01
Vacuum upholstered screens	0.01
TOILETS	
Change shower curtains	0.03
Clean sauna	0.09
Clean shower	0.08
Clean urinal	0.03
Clean washbasin	0.02
Clean wc	0.03
Clean wc cubicle	0.04
Descale sanitary fittings	0.02
Replenish supplies	0.01
REFUSE	
Clean external surface of wastebin	0.01
Clean internal surface of wastebin	0.01
Empty wastebin, clean internal and external surfaces and replace liner	0.05
Scrub wastebin	0.08

Task	Time/Unit (hours)
STAIRS	
Clean metalwork	0.05
Damp mop stairs	0.07
Dust control mop stairs	0.04
Dust fixtures and fittings of stairs	0.04
High dust on stairs	0.08
Pressure wash stairs	0.25
Scrub stairs	0.30
Scrub stair nosing	0.10
Spot clean stairs	0.03
Spot mop stairs	0.02
Spray extract stair carpet	0.50
Strip and redress stairs	1.60
Sweep stairs	0.05
Vacuum stairs	0.07
Vacuum stair edges and corners	0.04
Wet mop stairs	0.15
LIFTS AND ESCALATORS	
Clean escalator	0.30
Clean escalator sides	0.18
Clean lift (smooth floor)	0.10
Clean lift (inc. vacuum carpet)	0.08
Clean lift metalwork	0.50
Clean light fitting in lift	0.12
Scrub lift floor	0.10
Shampoo/spray extract lift carpet	0.12
Strip and redress lift floor	0.35
Vacuum lift carpet	0.02
Vacuum lift door channels	0.01
Wash lift walls	0.20
REFUSE	
Damp wipe wastebin	0.03
Empty and clean wastebin	0.04
Wash wastebins/ashbins	0.04
MISCELLANEOUS	
Clean drain	0.12
Clean kitchen appliance	0.10
Clean microwave	0.04
Clean refuse chute	0.06
Clean surfaces of kitchen appliance	0.08

Figure 10.3: Suitable pro forma for collecting data for unit counts and schedule of accommodation.

Figure 10.4: Partially completed pro forma for unit counts and schedule of accommodation at Neptune Stereophonics.

FLOOR LEVEL	ROOM NO.	ROOM DESCRIPTION	ROOM CLASSIFICATION	FLOOR TYPE	FLOOR AREA	LEVEL OF OCCUPANCY	SPECIAL REQUIREMENTS	Chairs	Blinds	Telephones	Blins	Glass doors	Ent. Mats	Other
GD	1	Reception	EN	c/tiles	160	MED		12		1	2	4	2	
	2	Disab. Toilet	TO	QT	21		Check Clean							
	3	Pantry	OF	V	9					1	1			
	4	Storeroom	SR	V	12						1			
	5	Conf. Rm	OF	C	124	MED		24up	6vl		1			
	6	Store	SR	V	17									
	7	Audio Vis.	CO	C	23		Hi Tech Mirror							
	8	Vestibule	CI	V	4									
	9	Storeroom	SR	V	8									
	10	Lift	CI	C	4									
	11	N.W. Stair	CI	V	81						1	1		
	12	Corridor	CI	V	45							2		
	13	Gents	TO	SF	18		Check Clean				1			
	14	Ladies	TO	SF	16		Check Clean				1			
	15	Open Plan of	OF	C	235	MED		10up	6vl	10	10			2us
	16	Corridor	CI	V	45									
	17	Plant Room	PR	V	16									
	18	Gents	TO	SF	18		Check clean							
	19	Ladies	TO	SF	16		Check clean							
	20	PABX	PR	V	10									
	21	Seminar Rm	OF	C	80	MED		16up	6vl	1	2			
	22	NE Stair	CI	V	74						1			
	23	Corridor	CI	V	45							2		

up = upholstered chair   vl = vertical blind   vn = venetian blind   us = upholstered screen

Figure 10.5: Pro forma survey analysis sheet

BUILDING:			TOTAL AREA:								
AREA		m^2	Up ch.	Tel.	Bin	Glass Door	Ent. Mat	Blind Vert	Blind Ven		

Figure 10.6: Partially completed survey analysis sheet for Neptune Stereophonics

BUILDING: NEPTUNE HQ		TOTAL AREA: 15295.5 m²							
AREA	m²	Up ch.	Tel.	Bin	Glass Door	Ent. Mat	Blind vert	Blind ven	Up. Screen
CARPETED OFFICES (OF)									
5. Conference Rm	124	24	1	1			6		
15 Open Plan	235	10	10	10			6		2
21 Seminar Rm	80	16	1	2			6		
27 Atrium	80	12	1	1	2				
36 Catering Of	6	1	1	1			2		
43 Security	60	6	5	6			4		
44 Office	48	5	3	5			4		
45 Office	4								
55 Open Plan	1933	44	44	48				16	18
61 Meeting Rm	116	7	7	7			6		
69 Office	481	27	26	28			12		
73 Office	182	10	9	9			6		
79 Office	45	5	5	4			4		
	3394	167	113	122	2		42	30	20
CARPETED STAIRCASE (CI)									
NW Stair	1* (81)								
NE Stair	1* (81)								
SE Stair	1* (81)								
SW Stair	1* (81)								
	4* (324)								

* Timed by unit count not area.

Figure 10.7: Suitable pro forma for the development of manpower requirements

BUILDING:		I			DENSITY: III	
ACCOMMODATION:		II			AREA: IV	
OPERATIONS	UNITS RATE	STAND- ARD TIME	HOURS OPER- ATION	ANNUAL FREQUENCY		TOTAL HOURS PER ANNUM
1.						
2.						
3.						
4.						
5.						
6.						
7.						
8.						
9.						
10. V	VI	VII	VIII	IX		X
11.						
12.						
13.						
14.						
15.						
16.						
17.						
18.						
19.						
20.						
				HOURS PER ANNUM		XI

Figure 10.8: Completed pro forma for the development of manpower requirements to clean carpeted office accommodation at Neptune Stereophonics

BUILDING: NEPTUNE HQ				DENSITY: MEDIUM	
ACCOMMODATION: OFFICES (CARPET)				AREA: 3394 m²	
OPERATIONS	UNITS RATE	STAND-ARD TIME	HOURS OPER-ATION	ANNUAL FREQUENCY	TOTAL HOURS PER ANNUM
1. Remove stains & spillages		0.03	1.02	260	265.2
2. Spot vacuum		0.06	2.04	260	530.4
3. Full vacuum		0.14	4.75	52	247.0
4. Spot wipe doors		0.03	1.02	52	53.0
5. Clean & polish glass		0.03	1.02	52	53.0
6. Spot wipe paintwork		0.03	1.02	52	53.0
7. Damp dust doors		0.06	2.04	12	24.5
8. Damp dust furniture		0.03	1.02	260	265.2
9. Damp wipe 'phones	113	0.02	2.26	52	117.5
10. Dust furn. frame*		0.20	6.79	52	353.1
11. Damp dust ledges etc		0.03	1.02	52	53.0
12. Spot clean up. chrs		0.03	1.02	52	53.0
13. Vac. up. chrs	167	0.01	1.67	12	20.0
14. Empty bins		0.03	1.02	260	265.2
15. Damp wipe int/ext of bins	122	0.03	3.66	12	43.9
16.					
17.					
18.					
19.					
20. *'Detail Dust'					
				HOURS PER ANNUM	2397.0

Figure 10.9: Annual frequencies for different service requirements

Service Requirement	Annual Frequency
Daily (5 days each week)	260
Daily (6 days each week)	312
Daily (7 days each week)	364
Twice weekly	104
Three times weekly	156
Four times weekly	208
Weekly	52
Every two weeks	26
Twice monthly	24
Four weekly	13
Monthly	12
Two monthly	6
Three monthly	3
Four monthly	3
Six monthly	2
Annually	1

11

DEPLOYING THE WORKFORCE

In Chapter 10 the procedure for calculating the manpower requirements has been discussed in detail. In this brief chapter we are concerned with the *deployment* of this labour to achieve the required manpower allocation.

The advantages and disadvantages of different shift patterns have been carefully considered in Chapter 6 in the context of fixing the times of cleaning. If a requirement has been set out in the *General Notes* to the Specification the contractor will have little flexibility when planning his labour deployment. If not, then he will have his own criteria for determining the logistics of offering a cleaning service. Some of these are considered below.

Scheduling is affected by a number of factors not the least of which is cost. In deciding how best to deploy the workforce to meet the manpower requirements therefore, it is necessary to consider all of these different aspects which have some bearing on the decision. These may include:

i. The occupancy status of the building i.e. whether or not the building is occupied at the time the cleaning is carried out. The absence of occupants leads to greater efficiency and hence higher productivity as cleaning staff are less prone to distraction or interruption.

ii. Whether there is a requirement for day cleaning activities. If certain tasks are to be carried out during the day time (such as check cleaning of toilets, cleaning dining rooms between meal and break times, or other janitorial type activities), then staff will have to be allocated during these hours.

iii. Holiday entitlement and expected levels of absenteeism.

iv. How it is intended to meet the requirements of the periodic schedules. For example periodic activities may be combined with core cleaning operations such that in a typical four hour shift, three hours of the cleaner's time may be allocated to tasks detailed in the core cleaning schedules and one hour may be spent carrying out periodic tasks. Alternatively periodics may be undertaken only as overtime. However if overtime rates are paid this will normally be too expensive. Subcontracting, or deploying a specific team who *only* undertake periodic tasks are other options which may be considered. Whatever procedure is finally selected cost will almost always be the overriding factor.

v. Preferred shift lengths, if not already specified in the *General Notes*. There are a number of factors to be taken into consideration. For example the building may only be accessible to cleaners at certain times of the day (and in some cases split shifts may be necessary, such that some cleaning is done early in the morning and the remainder is done early in the evening); there may be variations in occupancy throughout the year – as is the case with schools and colleges for example; public transport may be a problem after a certain time at night or before a certain time in the morning; short shifts might be selected because they reduce fatigue and carry lower national insurance contributions; long shifts might be preferred because 'down-time' is reduced, fewer people are required and hence overheads (including the provision of materials, equipment, uniforms, supervision etc.) will be lower.

Shift lengths may also affect absenteeism. Some will argue that the effects of absenteeism are diluted by the selection of short shift lengths. The logic to this argument is that if a job requires 8 hours of effort then if the shift length is 4 hours (i.e. 2 people are required) and one operative is absent then 50% of the workforce is missing whereas if four people each work 2 hours and one is missing then 25% of the workforce is absent and those that remain need only work 40 minutes overtime each to make up the shortfall. Others argue that short shifts engender absenteeism since there is less commitment to the job and less financial penalty results from being absent. The ability to recruit staff for short shifts may also be reduced since there is less financial benefit.

The question of supervision has also to be addressed.

Except on very small contracts the use of *working supervisors* i.e. supervisors who are also cleaning operatives, is not favoured. Either the effectiveness of their supervision will be diluted, or they will not be able to complete all of their cleaning duties with efficiency. Instead so-called *non-working supervisors* are preferred.

There are a number of points to consider:

i. The number of non-working supervisors that are required is dependent upon the size of the contract, not only in terms of labour (and hence monetary value) but in the context of its geographical size. A complex site with many buildings such as an airport or college will require a larger number of supervisors.

ii. There is an optimum number of personnel that only one supervisor can control. Think in terms of one supervisor to perhaps twelve operatives.

iii. The employment of every supervisor represents an overhead. Their wages, employment costs, uniforms etc. all add to the direct costs in operating the contract. Look upon this as money well spent since an effective supervisor will pay for themselves in the resultant increase in productivity of the personnel they directly control.

iv. Expect the supervisor to work a longer shift than other cleaning operatives. She needs to be on site a little before they arrive and may be required after they have left. She may also need to be involved in liaison with the client.

v. Decide the precise role of the supervisor. If the site also has a cleaning manager then the supervisor is more likely to spend most of her time in direct supervision. If however, there is no Site Manager she will also be involved in more varied administrative duties including purchasing.

Once all of these considerations have been taken into account it is possible to schedule the work in a manner which essentially meets the above requirements. There remains only a decision about the actual method of working. Options include allocating one specific area to each cleaning operative; allocating specific tasks to an operative or group of operatives such that there is a 'floor team', a carpet team, a Hi-Tech team etc; or setting up groups who move as a team from area to area throughout the building. Whichever is selected is usually decided by individual managers' own preferences.

How some of these factors influence the final costing of the cleaning operation will become clear in Chapter 12.

SUMMARY

In deploying the workforce to meet the time requirement a number of factors need to be considered. The alternatives may in any case be severely limited by external influences.

12

COSTING THE OPERATION

All that remains is the costing exercise.

The components of the contract price are aptly reviewed in Chapter 9 and are summarised by *Form 12* shown in Figure 9.12. The calculation of labour requirement has been studied in detail in Chapter 10 and some aspects of labour deployment are briefly examined in Chapter 11. Chapter 12 is intended to highlight some of the factors which need to be taken into account before the final cost can be derived.

Overview

Like any other commercial organisation a cleaning contractor will need to recoup all of his costs if he is to survive in business. Furthermore, to make survival worthwhile, he will expect to make a profit. His costs may be broken down into two components *fixed costs* and *variable costs*.

Typical fixed costs include the salary and employment costs of head office based management and administrative personnel (e.g. payroll clerks and secretaries) as well as accommodation costs which include rent or mortgage, and rates, and depreciation of fixed assets.

Variable costs tend to vary in proportion to the volume of business done by the organisation and will include both those costs which may be allocated directly to each specific contract as well as those which head office management choose to vary in ongoing budget reviews e.g. the cost of selling, advertising, marketing etc.

The price of a specific contract will therefore contain an element of *direct costs* i.e. those which relate only to that contract e.g. labour, materials and equipment costs; and *indirect* costs which are the organisation's non–contract–specific costs i.e. its fixed costs and that portion of its variable costs not related to the contract in question.

Let us suppose for example that Acme Industrial Cleaners Ltd., the incumbent contractor at Neptune Stereophonics have fixed costs of F, variable head office costs of V, direct costs related only to the cleaning of Neptune Stereophonics of D, and that they seek to make an annual profit of $P\%$ on the total price of the contract. Let us also assume that the monetary size of the Neptune Stereophonics contract is one nth of Acme's total turnover. Acme will expect the Neptune contract to recover its direct costs, make a *contribution* C to the general overhead, and make $P\%$ profit. Thus the total contract price T is given by the expression:

$$T = D + C + TP/100$$

In fact, if the contract is costed properly, then :

$$C = (F + V)/n$$

However, it may be that the contractor is keen to win the contract even though it may not make a full contribution to overheads and achieve its target for profit. In this case:

$$C < (F + V)/n \text{ and } P < TP/100$$

He may wish to do this because yours is a prestigious contract and he can gain some kudos by winning it. (Thereby helping him to win similar contracts). Indeed, in the early days of competitive tendering in the health service there were suggestions that some contractors had put in bids where $C = 0$ and $P = 0$ such that the total contract price represented only the direct costs of operation – this to enable them to claim that they were winning hospital contracts.

Clearly this condition cannot prevail for any significant length of time. Furthermore it is not in the best interests of the client to permit it to occur even though at first sight it may seem financially desirable to do so. For as has been discussed in a previous chapter, a contract which is not adequately costed will result either in financial difficulties for the contractor, or in his withdrawal of some labour in order to reduce the direct costs.

In the case of an in-house bid the apportionment of overheads may be more difficult and the calculation of C thus becomes somewhat notional. Contractors complain in some instances that this places them at a disadvantage in situations of competitive tendering against an in-house bid.

In practice, the value $(F + V)/n$ is often applied as a percentage of the direct costs. It is worth noting however that this calculation is only accurate *at the time the contract is priced*. If during the second year of the contract the contractor now has twice the volume of business in monetary terms (because his business is growing) then by definition his fixed costs will be unchanged, (although in practice he may have had to recruit extra head office staff to cope with the extra administration), his variable costs will not have doubled with the doubling of the volume of his business, but the value of n will have halved. His profit on this job is therefore increased. The following example illustrates the argument:

Suppose direct costs for Acme Industrial Cleaners current contract at Neptune Stereophonics are £100K in March 199x. Acme's fixed costs are £150K, their variable costs are £100K and they aim to make a profit of 5% on each contract. Suppose also the contract at Neptune Stereophonics represents one tenth of their annual turnover. Then substituting in the equation:

we get

$$T = D + (F + V)/n + TP/100$$

$$T = 100 + (150 + 100)/10 + T \times 5/100$$

$$T = 131.6K$$

of which £25K is a contribution to overheads and £6.6K is profit.

Suppose then by March 199x + 1 Acme's turnover has doubled such that the Neptune Stereophonics contract is now only one twentieth of their business. Suppose too that their fixed and variable costs have risen by say 30% to £195K and £130K respectively. If we now substitute in the equation to see what level of profit they are achieving from the Neptune contract, we obtain the following:

$$131.6 = 100 + (195 + 130)/20 + 131.6P/100$$

$$P = 11.7\%$$

The profit from the contract is thus almost double that expected at the time the price was originally determined.

Of course the converse may apply. If, in the above example, Acme's turnover had halved during the one year period instead of doubling, and if their fixed and variable costs had remained the same (as they often do in such circumstances) then the Neptune Stereophonics contract would have made a loss of £18.5K for Acme in the second year.

This simple theoretical explanation demonstrates the risks that cleaning contractors face when pricing a two or three year contract. It also explains in part, the acquisition trail upon which many contractors embarked during the nineteen eighties on the premise that increasing turnover would reduce overheads and therefore increase profitability of existing contracts or would allow keener pricing of future contracts. The premise was not always correct!

Direct Costs

Labour

The primary direct cost is labour and throughout Chapter 10 we have been concerned with the principles of manpower estimation. Only by precise estimation can we accurately *cost* the labour requirement. There are three aspects to consider:

a) <u>Minimum requirements to satisfy manpower calculations</u>

In Chapter 10, we calculated that the core cleaning of the office accommodation at Neptune Stereophonics required 2397.0 hours of effort per annum. Let us assume that 10600 hours are necessary each year to meet all of the requirements of the core cleaning schedules.

If a five day service is required, then 10600/260 or 40.76 hours of cleaning will be required each day.

Two important decisions must now be taken. These are:

i. How will the odd 0.76 hours be accommodated? In an entire year this represents almost 200 hours.

One possibility is that not all of the cleaners will work similar shift lengths. Another is that the daily requirement of 40.76 hours will be overallocated – with perhaps 12 cleaners working 4 hours, the residual 7.24 hours each week being allocated to periodic tasks and special operations.

ii. What is the preferred shift length? (If the General Notes dictate the shift length then the contractor will have no flexibility in this respect).

Clearly to achieve 42 hours per day for example, there are a number of simple alternatives – 14 cleaners working 3 hours, 12 working $3^1/_2$ hours, 7 working 6 hours etc.

Arguments concerning the merits and demerits of different shift lengths have been discussed earlier. However, at the costing stage there is a supplementary financial consideration. After a certain threshold level of pay, the employer becomes liable for national insurance contributions. Thus if operatives earn more than a given amount in any one week their employment costs will increase. The payroll cost will not simply be hours worked x hourly rate of pay but will become hours worked x hourly rate of pay + national insurance contribution. This will have a bearing on the total contract price. In the final analysis then, shift lengths may be determined using financial criteria alone, i.e. arranged to be of a length which avoids national insurance.

The complication of national insurance may also arise with overtime working.

Suppose for example an employee works 20 hours per week at an hourly rate of pay of £R. Assume that £20R lies just on the threshold for national insurance contributions to be activated. Assume then that the employee is asked to work 2 hours overtime and is paid for 3. The weekly wage now becomes £23R which exceeds the NI threshold. An additional payment of £N therefore becomes due. The hourly cost of the overtime is consequently not £1.5R as at first sight it may appear but (£3R + N)/2 per hour or in general terms, when the threshold is breached,

$$\text{hourly rate} = (mRh + N)/h$$

where m is the overtime multiplier (1.5 for time and a half, 2.0 for double time etc.), R is the normal hourly rate of pay, h is the number of hours overtime worked during the week, and N is the employers' national insurance contribution that applies.

For sub-threshold workers, this calculation becomes important to the contractor:

– when deciding how to allocate any shortfall in effort, either as a result of part hours occurring in his manpower calculations (the 0.76 hours in the example for Neptune Stereophonics).

– when deciding how to cover shortfalls due to absenteeism or sickness.

– when completing *Form 4b* of the Tender Return.

- when providing bank holiday cover since this may well inflate the individual's wage bill and thus activate the national insurance payment.

The costing for periodics follows these same principles.

b) <u>Holiday Cover</u>

The daily cleaning schedules must be met even when the operatives are on holiday. This means that the contractor must supply additional effort to cover these periods. There are several ways in which this might be achieved:

- staff can be employed on a temporary basis.
- existing staff can work overtime to complete tasks that would have been carried out by absent colleagues.
- on large sites, a pool of reserve labour may be on strength.

Unless operatives have no paid holiday entitlement all of the above options represent an additional direct cost.

The additional cost of temporary staff or a pool of reserve labour is obtained from the expression:

$$C = \Sigma \left[(d_1 h_1 R_1) + (d_2 h_2 R_2) \ldots\ldots + (d_n h_n R_n)\right]$$

If all staff have the same leave entitlement, same rate of pay and work the same hours each day, the expression may be simplified to:

$$C = n (dhR)$$

where C is the total annual cost, d is the number of days leave entitlement, h is the number of hours worked each day and R is the hourly rate of pay for each of n members of staff.

If cover is provided by means of *overtime* then the expression becomes:

$$C = m\Sigma \left[(d_1 h_1 R_1) + (d_2 h_2 R_2) \ldots\ldots + (d_n h_n R_n)\right]$$

where m is the overtime multiplier. In this case the arguments relating to national insurance contributions may also come into effect.

With experience a contractor will often apply the cost of holiday cover as a percentage of the calculated labour costs. Some cynics will argue however, that since turnover is so great in the contract cleaning industry, very few operatives will ever qualify for holiday entitlement and the contractor's allowance is a hidden profit. Others will also argue that the absence of cover during holiday periods will not be noticed since the work load is reduced because occupants of the building are themselves on holiday. This latter possibility can easily be avoided by effective monitoring.

c) Sickness and Absenteeism

The cost of absenteeism and sickness is usually derived historically and is applied as a percentage of the total labour cost. Of course this presumes that the contractor will make up any shortfall (as he is expected to do). If he does not, failure to provide absenteeism and sickness cover should result in a reduction in the total contract price, not an increase. Once again this can be controlled by effective monitoring.

d) Miscellaneous labour costs

Additional items for which allowance may have to be made are:

- maternity leave
- pensions
- bonuses
- a reduction in costs which may arise when as statutory sick pay reduces national insurance contributions.
- retainer payments, if the work is cyclical for example, as it may be in an academic institution.
- The implication of TUPE (The Transfer of Undertakings Protection of Employment Regulations) which, at the time of writing, still presents cause for confusion and concern.

In general most of the above factors can be accounted for by reference to historical data within the contractor's company.

Equipment

Equipment that is to be amortised over the life of the contract or partly depreciated over the life of the contract will be a direct cost to the contract.

The cost to the contract per annum for the equipment is given by the expression:

$$C = [(E - R) + M]/n$$

where C is the annual cost, E is the purchase price of the equipment or its net book value at the commencement of the contract, R is its residual book value at the end of the contract (which, if fully depreciated, will be zero). M is the total cost of maintenance (which will be an estimate based upon historical data and may be a percentage of the total cost of equipment) and n is the number of years for which the contract is to run.

Materials

For convenience many contractors express the cost of materials as a percentage of the labour costs. 4% is a typical figure.

Miscellaneous direct costs

Other direct costs may include the provision of protective clothing and the provision of transport for direct use in connection with the contract. When costing the provision of protective clothing it will also be necessary to allow for the cost of laundering (unless the employees are expected to launder their own uniforms).

If transport is required specifically for the contract, fuel costs and insurance will also need to be taken into account as will the vehicle excise licence and maintenance costs if the transport is bought rather than leased or hired.

Indirect Costs

Indirect costs as we have seen, include all the costs of head office support and the general costs of running a business. The apportionment of these amounts to any particular contract has been discussed in some detail at the beginning of this chapter.

However, it is worth bearing in mind that if every contractor 'calculates' the manpower requirement along similar lines then if wage rates are similar the direct costs in a contract will be similar for all those who tender. The competitive edge might therefore rest with those who have greatest control over their indirect costs. The discerning client may well recognise that the control of indirect costs is a sign of sound business practice and be persuaded accordingly.

Let us conclude then with an example that makes use of the above principles.

Neptune Stereophonic's European subsidiary Neptune Stereophonics (Ruritania) GmbH have modelled their specification on the one prepared by Roger Blackburn in the UK and are now going out to tender. With Ruritania's recent entry into the European Union there are no barriers to prevent British cleaning contractors from competing with local Ruritanian companies and Acme Industrial Cleaners decide to bid for the contract. Their representative was present at the site visit and using the specification and measurements provided, they have calculated a manpower requirement of 11700 hours over 52 weeks to meet the 260 day core cleaning requirements and a further 4200 hours for the periodics and special operations. Neptune Stereophonics (Ruritania) GmbH have placed no restrictions on the hours during which cleaners may be on site but have stipulated that the plant will operate during each of the eight Ruritanian bank holidays and will therefore require a cleaning service on those days. Acme Industrial Cleaners Ltd. have contacted the local Ruritanian Consulate in Chorlton–cum–Hardy and have established that typical pay rates for cleaners in Ruritania are 3.00 Ruritanian Zlotys (the symbol for which is the British £ sign to symbolise the close links between the House of Windsor and the Zendan royal family). Supervisors are typically earning 4.00 zlotys per hour (£4.00). Employees of Ruritanian labour are required to pay national insurance contributions of 4.6% of the weekly wage wherever that employee's weekly wage exceeds 61.00 zlotys (£61.00). Market research has also established that overtime working and bank holiday working normally attract double time except for time worked on the important festival of Count Dracula's birthday, celebrated each year on 24 March when triple time has to be paid. Acme have also discovered that because of the shortage of manual labour in Ruritania, sickness, holidays and absenteeism will need to be covered by overtime rather than by use of temporary staff. They have also been told by the Ruritanian Consulate that all employees must by law be allowed 10 days paid holiday each year (in addition to Bank Holidays).

Maternity leave is not allowed in Ruritania. Although Ruritania have recently entered the European Union, they have not joined the Exchange Rate Mechanism and at the present time £1.00 = £1.00 (i.e. 1 Ruritanian Zloty = £1.00 Sterling). Acme have decided, in this their first attempt to win business overseas, to reduce their profits to 6%. They anticipate an allowance of 5% for absenteeism and sickness. They are seeking a contribution to overheads of £25K Sterling from the contract.

Their calculations were as follows:–

Manpower requirement to complete
core cleaning schedules   =   11700 hours per 52 weeks

                                =   11700/260

                                =   45 hours per day

This can be achieved in several ways:

I.   using 9 cleaners each working 5 hours per day

      i.e. 25 hours per week per cleaner

II.   using 10 cleaners working 4.5 hours per day i.e. 22.5 hours per week per cleaner

III. using 12 cleaners working 4 hours per day i.e. 20 hours per week per cleaner with 3 hours or surplus effort available each day to contribute to periodic operations

IV.   using 15 cleaners working 3 hours per day i.e. 15 hours per week per cleaner

Other variations may also be considered.

Acme expect to use 1 supervisor who will work 30 minutes longer than the shifts worked by operatives.

In each of the above options, the weekly wage for each cleaner is given by the expression:

$$W = h \times R$$

where W is the wage, h is the number of hours worked per week and R is the hourly wage rate. Thus the wages under each of the above options is:

I	25.0 x 3.00	=	£75.00 per week
II.	22.5 x 3.00	=	£67.50 per week
III.	20.0 x 3.00	=	£60.00 per week
IV.	15.0 x 3.00	=	£45.00 per week

However options I and II become liable for payment of national insurance contributions at the rate of 4.6% of the weekly wage. Thus the annual labour cost (excluding supervision), in general terms is:

$$C = (W + I) \times N \times 52.143$$

where C is the cost, W is the wage I is the national insurance (given by $Wr/100$ where r is the rate (%) once the threshold is breached) and N is the number of cleaners employed, or:

$$C = (W + Wr/100) \times N \times 52.143$$

52.143 being the actual number of weeks in a year.

So, for each of our options the annual labour cost is:

I. $C = (75 + 75 \times 4.6/100) \times 9 \times 52.143$

       $= £36815.57$

II. $C = (67.5 + 67.5 \times 4.6/100) \times 10 \times 52.143$

       $= £36815.57$

III. $C = (60 + 60 \times 0/100) \times 12 \times 52.143$

       $= £37542.96$

However in this option 15 surplus hours per week are available for periodics. This represents a value of $15 \times 3 \times 52.143 = £2346.43$. Thus the actual cost of the core cleaning operations is:

$$£37542.96 - £2346.43 = £35196.53$$

IV.   C = (45 + 45 x 0/100) x 15 x 52.143

      = £35196.53

It becomes clear from these calculations that options I and II cost Acme the same and options III and IV cost Acme the same. However the financial benefit to not exceeding the national insurance threshold is £1619 per annum. (£36815.57 − £35196.53). £4857 over a three year contract.

At this stage however, no allowance has been made for absenteeism, sickness and holidays.

Acme expect to allocate 5% for absenteeism, sickness and holidays. They need make no provision for maternity leave since it is not allowed in Ruritania.

5% of the annual manpower requirement is (5 x 11700)/100 = 585 hours per annum, which will be paid at double time. In effect this increases the wages bill by 585.0 x 2 x £3.00 = £3510.00 per annum or £67.31 per week. If this £67.31 is spread equally over the 15 cleaners of Option IV, their average weekly wage will rise by £4.49 to £49.49 which remains below the national insurance threshold. If however it is spread over the 12 cleaners of option III, the calculation changes. Firstly there are 15 hours spare each week on the proposed shift pattern for option III, at present held in reserve to support the periodics effort. This represents 782.15 hours per annum. Thus, on this shift pattern there is sufficient slack to accommodate a shortfall in effort of 585 hours due to absenteeism and sickness without the need to pay overtime.

Allowing for these charges, the cost of option III and IV therefore become:

III   C = (60 + 60 x 0/100) x 12 x 52.143

      = £37542.96

This is the same as before except that of this, £649.95 (197.15 hours) of effort is still available for periodics. So the core cleaning cost is £36893.01.

IV   C = (49.49 + 49.49 x 0/100) x 15 x 52.143

     = £38708.36

Option III has thus become the most cost effective.

But we have yet to allow for holidays and bank holidays and because of the shortage of casual labour in Rurutania, these must be worked as overtime.

The additional requirement in the case of Option IV may be calculated as follows:

10 days Annual Leave of 3 hours per day
for each of 15 cleaners, covered at double time   =   900 hours

8 days Bank Holidays paid at double time except
Count Dracula's birthday, paid at triple time = 9 extra* days
at single time = 9 x 3 x 15
                                                        =   405 hours
                                                            ———
                                                        = 1305 hours

These 1305 hours increase the wages bill by £3915.00 per annum or £75.08 per week. Spread equally over 15 cleaners their average weekly wage will rise by £5.00 to £54.49.   Still below the national insurance threshold.

Thus the cost of option IV allowing for sickness, absenteeism and holidays, becomes:

$$C = (54.49 + 54.49 \times 0/100) \times 15 \times 52.143$$

$$= £42619.08$$

If the holiday and bank holiday cover are allocated across the 12 cleaners of Option III however, the calculation changes significantly.

We have seen, after applying sickness and absenteeism allowances, they are still carrying a residual 197.15 hours provisionally allocated to periodics. Thus the 1305 hours for holiday cover can be reduced to 1305.00 – 197.15 = 1107.85 hours. This increases the wages bill by £3323.55 per annum or £63.74 per week. Amongst 12 cleaners this becomes an extra £5.31 per cleaner, thereby increasing their weekly wage to £65.31. This now exceeds the national insurance threshold and the cost of Option II becomes:

$$C = (65.31 + 65 \times 4.6/100) \times 12 \times 52.143$$

$$= £42745.33$$

Option IV is thus £243.57 cheaper than Option III.   £730.70 over the life of the contract.

Acme finally know how to staff the contract and in order to complete the tender return shown in Figure 9.12 as *Form 12* they can supply the following information.

i.   Total weekly labour hours

   $W = n \times h = 15 \times 15$ =   225 hours

---

* 7 days at double time and 1 day at triple time is 17 days. However in assuming 1 year = 260 days, in the basic calculation, 8 of these days have already been costed.

ii. Total weekly cost of cleaning

$C = n \times h \times R = 15 \times 15 \times 3.00$ = £675.00

iii. Weekly cost of supervision

$S = s \times h \times R_s + s \times h \times R_s \times 4.6/100$

$= 1 \times 17.5 \times 4.00 + 1 \times 17.5 \times 4.00 \times 4.6/100$ = 73.22

iv. Annual labour cost

$L = (C + S) \times 52.143$

$= (675 + 73.22) \times 52.143$ = 3904.44

v. Bank holiday costs

$B = b \times n \times h \times R$ = 1215.00

$= 9 \times 15 \times 3 \times 3.00$

vi. Holiday costs

$H = mn (d \times h \times R)$

$= 2 \times 15 (10 \times 3 \times 3.00)$ = 2700.00

vii. Allowance for sickness and absenteeism

$A = m (p \times T)/100$

$A = m (p \times T \times R)/100$

$= 2 (5 \times 11700 \times 3.00)/100$ = 3510.00

Where:

n is the number of cleaners

h is the hours worked per week by each cleaner/supervisor

R is the operative wage rate

S is the number of supervisors

R_s is the supervisor wage rate

b is the number of days bank holiday not covered by the original scheduling

m is the overtime multiplier

p is the percentage allowance for sickness and absenteeism

T is the total manpower requirement before all allowances (excluding supervision)

A similar series of calculations may be carried out when costing periodics except *Form 9* of the tender return asks for periodics to be costed separately. These costings should, on a job by job basis, be uplifted with allowances for sickness, absenteeism and holidays.

To simplify future calculations in this example let us assume that the total cost of periodics and special operations is £20K per annum.

The direct costs that remain are materials, equipment and protective clothing. Additionally Acme have to make some allowance for head office support i.e. general overheads. Let us assume the cost allowed for materials is £4,000. Let us also assume that equipment, (with maintenance), costs will be £10,000. As the contract is being negotiated over two years, Acme have several options.

1. They may choose to depreciate the equipment over the life of the contract i.e. at the rate of £5K p.a.

2. They may plan straight-line depreciation of the equipment over four years (i.e. 25% p.a.) in which case it will have a book value at the end of the contract of £5K. Thus the annual cost to the contract will be £2.5K.

3. They may use second hand equipment brought in from a previous 2 year contract which has now finished. In this case they may fully depreciate its book value over the life of the new contract.

As Ruritania's National Grid operates on the novel voltage of 127.3V and 46.4 Hz Acme will have to use new equipment, purchased in Ruritania, which will be useless at the end of the contract. The cost of £10K will therefore need to be depreciated over the life of the contract.

Each operative and the supervisor will need two overalls. Thus a total of 32 overalls will be required. These cost £20 each and are laundered every week at £1.00 per overall. Thus the total protective clothing cost is:

Purchase of 32 overalls at £20	=	£640.00
Laundry of 16 overalls for 104 weeks at £1.00	=	£1664.00
		£2304.00

or £1152.00 per annum

Acme are seeking a contribution to overheads of £25K from the contract and a profit of 6% of the total contract price.

From information derived in this example we find that Acme's direct costs are:

$$D = L + Pe + B + H + A + E + M + U$$

where D are the direct costs

L is the annual cost of all labour

Pe is the annual cost of periodics

B is the annual cost of bank holiday cover

H is the annual cost of holiday cover

A is the annual allowance for sickness and absenteeism

E is the annual cost of equipment

M is the annual cost of materials

U is the annual cost of uniforms

then D = 39014.44 + 20000.00 + 1215.00 + 2700.00 + 3510.00 + 5000.00 + 4000.00 + 1152.00

= £76591.44

So by substitution in the equation $T = D + C + TP/100$ given on page 185, we find

T   = 76591.44 + 25000 + T x 6/100

T   = 101591.44 + 6T/100

94T = 101591.44

T   = £108076.58 per annum

Footnote:

Acme's bid to clean Neptune Stereophonics (Ruritania) GmbH headquarters for £110K in the first year and £115.5K in the second (allowing 5% for inflation) was rejected in favour of a local Ruritanian contract cleaner Rupert of Hentzau Cleaners AG.

SUMMARY

The ultimate contract price depends upon a combination of direct costs, overheads and profit.

Direct costs are primarily accounted for as labour costs but also include materials and equipment.

Thoughtful deployment of labour may obviate the need to pay national insurance contributions resulting in a more competitive bid. However there are arguments against short shift patterns even though they may avoid such national insurance contributions. Other hidden employment costs including sickness, holiday and pension payments should not be overlooked.

Indirect costs should be minimised by good management although each contract will need to make a smaller contribution to overheads as the volume of business increases.

A year on year increase in the volume of business leads to increased profitability within the life of the contract.